# TARANTULAS: KEEPING AND BREEDING ARACHNIDS IN CAPTIVITY

KATHLEEN HANCOCK AND JOHN HANCOCK

R&A

1992

Published by R & A Publishing Limited
Distributed by R & A Research & Information Ltd
  2nd Floor
  24 Belvedere Road
  Taunton, Somerset
  England TA1 1HD.

Distributed in The Americas by
  Eric Thiss
  Serpent's Tale
  464 Second Street
  Excelsior, MN 55331
  USA

© text: R & A Publishing Limited
© photographs: K. Hancock and J. Hancock, except
    front cover, Plate 4 a, b, c and f. © Roy Dunn.

All rights reserved. No part of this publication may be reproduced, stored in a retrieval system or transmitted, in any form or by any means, electronic, mechanical, photocopying, recording or otherwise, without the written permission of the copyright owners.

ISBN 1 872688 06 3

Front cover: *Lycosa carolinensis*, a Wolf Spider from Arizona

Cover design: KRG Welch

Photoset, printed and bound in Great Britain by
The Longdunn Press Ltd, Bristol

# Contents

| | |
|---|---|
| Preface | vi |
| Introduction | 1 |
| Chapter 1: What is a tarantula? | 5 |
| Chapter 2: Anatomy, Classification and Identification | 9 |
| Chapter 3: Choosing a Tarantula | 20 |
| Chapter 4: Housing | 24 |
| Chapter 5: Feeding | 41 |
| Chapter 6: Moulting | 53 |
| Chapter 7: Sex Determination | 58 |
| Chapter 8: Breeding | 67 |
| Chapter 9: Rearing Spiderlings | 71 |
| Chapter 10: Atypidae, Dipluridae and Ctenizidae | 75 |
| Chapter 11: Araneomorphae - The True Spiders | 81 |
| Chapter 12: Scorpions | 93 |
| Chapter 13: Other Arachnids | 101 |
| Chapter 14: Collecting and Conservation | 111 |
| Species File | 117 |
| Glossary | 138 |
| References and Further Reading | 141 |
| Index | 143 |

# Figures

Fig. 1 Cave drawing in Castellon, Spain.
Fig. 2 Inca spider at Nazca.
Fig. 3 Dorsal view of a Theraphosid spider.
Fig. 4 Ventral view of a Theraphosid spider.

Fig. 5 Leg segments.
Fig. 6 Chelicera showing fang and poison gland.
Fig. 7 Fangs of Mygalomorph and Araneomorph spiders.
Fig. 8 Ocular arrangement.
Fig. 9 Foveal groove.
Fig. 10 A burrow of a Theraphosid spider.
Fig. 11 A 'scrape' under a rock.
Fig. 12 An arboreal web.
Fig. 13 A range of plastic containers suitable for spiderlings to adults.
Fig. 14 Light bulb heating.
Fig. 15 A heating pad.
Fig. 16 A heating cable.
Fig. 17 Housing constructed from a building block. *Haplopelma minax* in burrow.
Fig. 18 *Lasiodora parahybana* in an aquarium.
Fig. 19 Juvenile *Euathlus smithi* in an aquarium.
Fig. 20 A custom made container for *Pterinopelma saltator* (The White Collar).
Fig. 21 Feeding record.
Fig. 22 A holding jar for locusts.
Fig. 23 Locust breeding set-up.
Fig. 24 A pooter.
Fig. 25 Predicting an approaching moult.
Fig. 26 The process of ecdysis or moulting.
Fig. 27 External sex determination.
Fig. 28 Male Theraphosid sex organs.
Fig. 29 Female Theraphosid sex organs.
Fig. 30 Female epigynum showing spermathecae (viewed from the inside).
Fig. 31 The two basic forms of spermathecae.
Fig. 32 Side view of the female's epigynum area.
Fig. 33 Male epigynum area without and with accessory organs.
Fig. 34 Container to house a spiderling.
Fig. 35 Spiderling rearing containers.
Fig. 36 Purse Web spider in its tube.
Fig. 37 Trap door spider in its burrow.
Fig. 38 Scaffold web.
Fig. 39 *Nephila senegalensis*.
Fig. 40 *Nephila* in a frame.
Fig. 41 Web of *Argiope bruennichi* showing stabilimentum.
Fig. 42 Egg-sac of *Argiope bruennichi*.
Fig. 43 A safe method of collecting scorpions.
Fig. 44 *Pandinus imperator*.
Fig. 45 *Scorpio maurus*. The Large Clawed Yellow Scorpion.

Fig. 46 Building block housing carved for Tiny Malaysian Wood Scorpions.
Fig. 47 Kenyan Orange Scorpion (Buthidae).
Fig. 48 A classification of Arachnids.
Fig. 49 Amblypygi.
Fig. 50 Distribution of Amblypygi.
Fig. 51 Uropygi.
Fig. 52 Distribution of Uropygi.
Fig. 53 A Chilean Laniotore and a Harvestman.
Fig. 54 A Red Velvet Mite.
Fig. 55 A pseudoscorpion.
Fig. 56 A solifugid.
Fig. 57 Distribution of Solifugids.

# Preface

During the past few years there has been a growing interest in the keeping of tarantulas and other arachnids. Unfortunately there has been little information available resulting in the untimely demise of many of these creatures and disappointment for the keeper. They are rather exacting in their requirements except for a few species which survived despite the conditions in which they were kept. With the availability of a greater number of species from a variety of habitats more data is needed to be able to keep them successfully. The information in this book is the result of keeping and successfully breeding tarantulas for a number of years. The methods described have been found to be the most suitable. We had disasters when we first began to keep tarantulas and this was extremely frustrating as there did not appear to be any reliable information readily available. Through trial and error we got it right and now have around 4500 tarantulas of 80 different species at various stages of development. The breeding programme which we have worked out has resulted in many fertile egg-sacs and most of our spiders are captive bred. Each one has a code number and accurate records are kept so it is possible to work out a 'pedigree' for each one. By doing this we are, hopefully, relieving the pressure on the wild stock and the knowledge we have gained can be passed on.

As the number of enthusiasts has grown, so has the success of the British Tarantula Society which has members world wide. This is a friendly society with members always willing to help each other. Through the society many captive bred spiderlings become available and mature males are used to their full potential. There is a quarterly journal for members and an annual show. Many members may only keep one spider as a pet, others have expanded their collections. As well as the general care of tarantulas, there is a more scientific aspect to the society with work being done on classification and identification. There are groups around the country which hold regular meetings for members.

For those of a more scientific bent, the British Arachnological Society produces a bulletin of scientific papers three times a year as well as a newsletter. They hold field trips within Great Britain, but their work also covers Mygalomorphs, and they also have members throughout the world.

With the growing awareness for the need for conservation the more knowledge that is gained and passed on the better. Once a species becomes extinct, it is gone for ever. Tarantula keepers can help the situation by keeping captive bred spiders, reducing imports to a minimum.

We have a considerable amount of pleasure from our spiders and other assorted arachnids and hopefully some of the knowledge we have gained will be of use to others.

Kathleen and John Hancock<br>
Southminster, Essex, 1991

# Introduction

Compared to spiders, man is a relative newcomer to our planet. Modern man appeared a mere 40,000 years ago whereas there is fossil evidence from the Devonian period, 395 million years ago, that spiders existed then. Perfect specimens of spiders have been found preserved in amber formed during the Oligocene period, 38 million years ago, around the Baltic coast.

It is thought that spiders evolved from aquatic species that emerged from the primeval oceans at the very dawn of life on dry land. Scorpions, which, like spiders, are members of the class Arachnida, are thought to be descended from the extinct eurypterids or sea scorpions which resembled giant lobsters. However the ancestry of spiders is not so positive because fossil evidence is incomplete. The evidence that has been found shows that over millions of years some species have remained virtually unchanged. Like the crocodile, they were so suited to their environment that evolutionary adaptations were unnecessary for their successful survival.

Man has been aware of the presence of spiders for many thousands of years. The earliest known representation is a drawing on a cave wall in Gasulla Gorge, Castellon, Spain. It depicts a spider and flies. Through the centuries, on every continent, there have been legends and superstitions surrounding spiders. Fig. 1.

The word ARACHNOLOGY is derived from the Greek word ARACHNE, meaning spider. In mythology, Arachne was a skilled weaver who boasted that she could weave more beautifully than Athena, the Goddess of Arts and Crafts. Athena, disguised as an old woman, warned Arachne not to boast. Arachne, however scorned her advice and Athena revealed herself and accepted the challenge. Athena wove a beautiful tapestry depicting mortals being punished for their pride; Arachne's depicted the

Fig. 1 Cave drawing in Castellon, Spain.

misbehaviour of the Gods and Goddesses and was even more beautiful. Athena ripped the fabric when she saw that it was better than her own. In terror, Arachne attempted to hang herself. Athena took pity on her and transformed her into a spider. Arachne's skill survived in spinning webs.

A puzzle left to us by the ancient Incas some 2000 years ago are the lines of Nazca in Peru. These consist of lines and drawings on the ground which are so large that they only became evident with the advent of flight. One of these drawings depicts a gigantic spider about 50 metres across. The significance of the lines has been attributed to an ancient form of astronomy, but the drawings surely depict creatures considered to be of some importance to the Incas. Fig. 2.

Dr. Thomas Mouffet, father of the nursery rhyme character Little Miss Mouffet and author of the first insect book in English, showed a keen interest in spiders. He considered the webs of the House Spider (*Tegenaria domestica*) as decoration. Little wonder his daughter was frightened of spiders – he dosed her with them to cure many ailments! Dr. Mouffet was not alone in the belief that spiders cured ills. References have been made to them being swallowed to cure fevers, from the sixteenth to the nineteenth centuries in Britain. This could be done whole, wrapped in a raisin or spread on bread and butter! The belief had been that a spider in a bag or nutshell hung round the neck would ward off fever. Perhaps by swallowing the charm the benefits would be increased.

The folk-lore of many countries tells of a local hero who escaped pursuers through a spider building a web over the entrance to a hiding

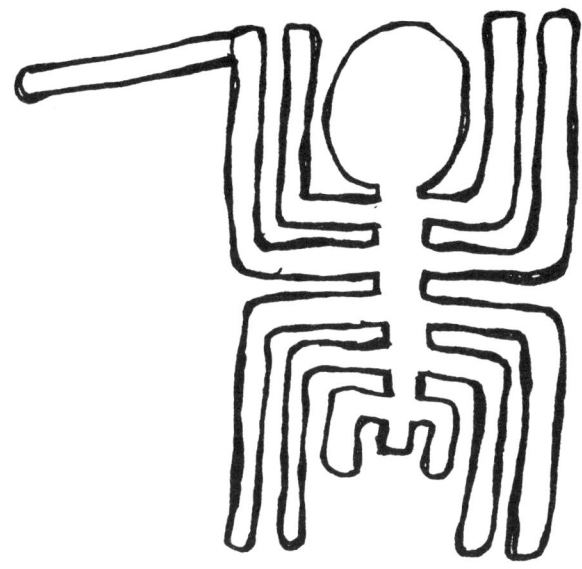

Fig. 2 Inca spider at Nazca.

place. David is said to have escaped from Saul in this way, as did Mohammed from the Corieshites. There are similar tales from Italy, Japan and Bulgaria. The perseverance of the spider is said to have inspired Robert Bruce – if at first you don't succeed, try, try again.

Spiders are considered by many to be lucky. The Money Spider, a small black Linyphiid, is said to bring money or a gift. It is also considered unlucky to kill a spider. Customs in a variety of countries support the idea of spiders bringing good luck. The appearance of a spider in the path of a bride or groom on their way to their wedding is supposed to indicate future happiness and prosperity. Hindus in Eastern Bengal collect spiders to release at weddings and in Egypt a spider is put in the bed of a married couple on their wedding night. Spider divination has been, and still is, popular in West Africa. In western Cameroon a Baboon spider, *Heteroscodra crassipes*, which lives in burrows, is believed to be in touch with the sacred underworld where the spirits of ancestors live. Less than a century ago the penalty for deliberately killing one of these spiders was death. In the Kalahari Desert of south west Africa the Bushmen collect Baboon spiders by irritating them with a stalk of grass until they leave their burrows. They are then placed on a skewer over a fire to singe the hairs. This is part of the initiation rite of young men before their first hunt. The sport of Baboon baiting was enjoyed by young boys of the Zulu and Matabele tribes as a pastime whilst guarding their herds. The spiders were placed within a circle of stones and bets placed, a similar sport to cock fighting.

So, over the ages and throughout the world, spiders have played their part in human culture. Could this be due to their amazing ability to spin web or their predatory lifestyle? Whatever the superstitions of the past, spiders play an important part in the ecosystem. The very fact that they have remained unchanged for so long proves the success of their strategy.

Chapter One

# What is a tarantula?

These giants of the spider world are known by a variety of common names. 'Tarantula' is a misnomer. There is only one true tarantula in the world and this is relatively small. It is a Lycosid, or Wolf spider, found in southern Europe. It is easy to see how this name came to be used when referring to large, fearsome spiders as its bite was thought to cause a type of madness which could only be cured by partaking in a frenzied dance, the Tarantella. This dance took its name from the town of Taranto in northern Italy. Hence the name 'tarantula' seemed appropriate to large, apparently dangerous spiders. It is the most frequently used common name for these creatures which are also known as Banana spiders and Bird-eating spiders. Neither of these names is strictly accurate. Banana spider is used because these large spiders were often found in bunches of bananas imported from tropical regions. It is more likely that these spiders were *Heteropoda venatoria*, a giant Huntsman spider. It is now extremely rare to find spiders or any other insects in imported bananas as these are treated to remove any such pests. The fact that some species of large spider live in trees could be the reason for them being given the name Bird-eater. No doubt they do take small birds occasionally in the wild, but these are by no means their staple diet. However it does sound more impressive than 'Insect-eater'! In Africa these large spiders are known as Baboon Spiders. The name originated in Equatorial Africa, probably indicating respect for these long-lived hairy occupants of deep burrows. Most Baboon spiders are very aggressive.

Spiders and all other arachnids in this book belong to the Class Arachnida. This Class is divided into Orders. The Order to which all spiders belong is Araneae. The Order Araneae is further divided into three Sub-orders which are based on the evolutionary development of their members. The first Sub-order is Mesothelae, primitive segmented spiders which are now only found in small areas of eastern Asia. The second Sub-order, Orthognatha (the Mygalomorph spiders), is the one which includes tarantulas. The third Sub-order is Labidognatha, the Araneomorphs or True spiders.

The Mygalomorphs are divided into fifteen families which include Theraphosidae, Dipluridae, Ctenizidae and Atypidae. It is to these four families that commonly kept tarantulas belong. So to be strictly accurate a 'tarantula' is a Mygalomorph and the most popular are Theraphosids.

In terms of evolution, tarantulas are more primitive than the true spiders with regard to anatomy, the main difference being their fangs. Those of a true spider move parallel to the body, i.e. an Araneomorph can clasp its prey. The Mygalomorph's fangs fold under the body. It must strike downwards, against the ground, for example, to impale its prey. The second main difference is the fact that Mygalomorphs have four book-lungs whereas the more advanced true spiders have only two. This will be dealt with further in the chapter on Anatomy.

Although tarantulas are thought of as large, hairy spiders, they vary in size from having a body length of less than a millimetre to the giants of the spider world which have a leg-span of 25 centimetres. Out of the world's 100,000 or so named species of spider, more than 650 species of Theraphosid have been described and named. It is believed that there are many more so far unknown species inhabiting the rain forests. When a 'new' species is discovered, it must first be established that it has not already been described and classified. Much of this work was done in the late nineteenth and early twentieth centuries. A scientific paper is then produced and the spider is given a name. For the purposes of accuracy scientific names are derived from Greek and Latin. This avoids the confusion which often arises if common names are used, particularly if the spiders are to be used for breeding.

Of the Mygalomorphs, the only one found in Great Britain is of the family Atypidae, the Purse-web spiders. These spiders live in a burrow which is extended by a closed silken tube, inside which the spider spends its life. It catches its prey by spearing it through the tube's silken wall. It then makes a hole through which it pulls the prey, then it reseals the tube. The British species is *Atypus affinis* and one of the American species is *Atypus bicolor*.

The family Ctenizidae is made up of those spiders that live in a burrow and seal the entrance with a trap-door, hence the common name Trap-door Spiders. This lid is hinged and the spider lies in wait just inside the burrow. When prey passes it grabs the unsuspecting insect, retreats into its burrow and closes the lid. One of the characteristics of these spiders is the presence of a rastellum on the chelicerae which facilitates burrowing.

Dipluridae is the family of spiders which construct sheet webs on which they live. They have particularly long spinnerets. Although *Atrax robustus*, the notorious Sydney Funnel-web, belongs to this family, not all the other species have such potent venom. Dipluridae are interesting spiders to keep as they are much more active than Theraphosidae, although not very large. The largest European species is *Macrothele calpeiana* which is found in southern Spain. This is an attractive spider which is quite easy to keep in captivity.

Theraphosidae is the family to which the largest tarantulas belong. The best known is undoubtedly the Mexican Red Knee (*Euathlus smithi*) which was exported in vast numbers from its natural habitat for the pet trade. This is a docile spider which can be tolerant of less than ideal conditions but because of its popularity it has been over-collected in many areas. The temperament of tarantulas varies greatly, not only from species to species, but from one individual to another of the same species. For this reason they should always be treated with respect, particularly species from the 'Old World'. African and Asian species are quick to show aggression, whereas those from the 'New World' can give warning of displeasure or fear. The abdomen of these spiders is covered with urticating hairs. When the spider is frightened or annoyed it 'throws' these hairs using its back legs. At this point it should be left well alone as the hairs can be irritating and often produce a rash. As for the spider, repeated throwing of hairs will result in a bald patch on the abdomen. The use of these hairs does not mean that the spider will not bite if aggravated. If bitten, the symptoms are generally mild unless the victim is hypersensitive. (Some people are allergic to bee or wasp stings.)

Tarantulas are mostly found in tropical and sub-tropical latitudes and occupy a variety of habitats from semi desert to tropical rain forest. They construct burrows or make tube webs high in trees. Species from the Americas are most commonly found in pet shops, although many of these countries are restricting the export of their native fauna. More species are being imported into the United Kingdom from Africa, but these Baboon spiders do tend to be aggressive, as do species from Asia. The natural habitat of each species must be carefully considered when keeping them in captivity as the more realistic the conditions, the happier and healthier the spider will be. The requirements of different species will be dealt with in more detail in the 'Species File'.

To sum up, a tarantula belongs to the Sub-order Mygalomorphae and the majority of commonly kept species are Theraphosidae. They have downward striking fangs and two pairs of book-lungs and it is by these two characteristics that Mygalomorphae and Araneomorphae can be separated.

Chapter Two

# Anatomy, Classification and Identification

**Anatomy** The body of a Theraphosid spider (tarantula), as with the Araneomorphs (true spiders), is divided into two distinct parts. The 'head', or anterior prosoma carries the legs, eyes, chelicerea, pedipalps, fangs, etc. and the abdomen, or posterior opisthosoma, carries the booklungs, spinnerets and the opening of the reproductive organs. (Figs. 3 and 4)

Spiders' bodies have evolved from being segmented. The six segments that make up the head and thorax have fused together to form the cephalothorax. The dorsal area of the cephalothorax is composed of a toughened shield, the carapace. This carries the eyes, generally eight but in some Araneomorphs, six. The underside of the cephalothorax is strengthened by two further plates, the sternum and the much smaller labium which covers the mouth parts. Six pairs of appendages are attached to the cephalothorax. As well as the four pairs of legs there are the pedipalps and chelicerae. The chelicerae are used for catching and crushing prey and end with the fangs. The pedipalps are also used for manipulating food, but in the mature male they are used as sexual organs to insert the sperm into the female. The walking legs illustrate one of the characteristics of the Arthropods – the jointed limb. The flexible joints are made possible by a thin joint membrane. This membrane also joins the three plates together and it is the splitting of this in a prescribed place and sequence that allows the spider to moult.

The four pairs of walking legs have seven segments, beginning from the body 1. the coxa, 2. the trochanter, 3. the femur, 4. the patella, 5. the tibia, 6. the metatarsus and 7. the tarsus. (Fig. 5). The pedipalps have six segments, similar to the legs but without the metatarsus. The front two pairs of legs are relatively long and the first pair in particular is often used

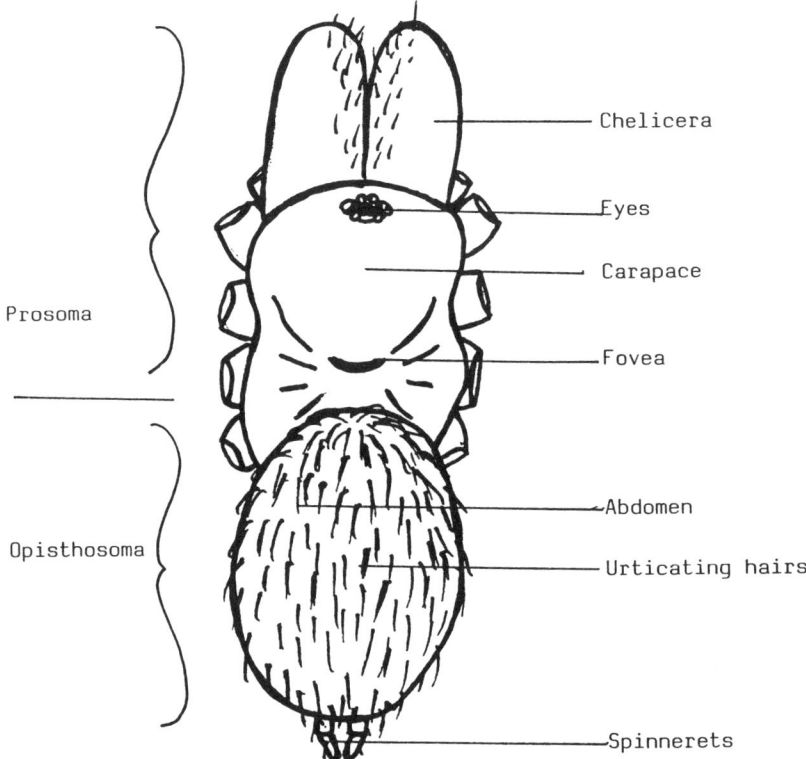

Fig. 3 Dorsal view of a Theraphosid spider.

as feelers to probe the environment. Sensory hairs densely cover the legs and these are very sensitive to movement, even air currents. At the tip of the tarsus are two bent claws which are generally serrated. These are the reason for it being unwise to handle tarantulas as the claws can become entangled in fabric and the spider may throw off a leg. From directly under the claws, the underside of the tarsus and metatarsus are often covered by scopulae (hairs). The amount varies according to species. These scopulae enable the tarantula to walk easily on smooth vertical surfaces such as the glass sides of the container. The number of contact points is increased greatly as each scopula hair divides into thousands of extensions rather like a broom.

    The chelicerae are the first appendages on the prosoma. (Fig. 6). Each chelicera consists of two parts – a thick basal segment and a movable articulated fang. Generally both sides of the groove in which the fang rests

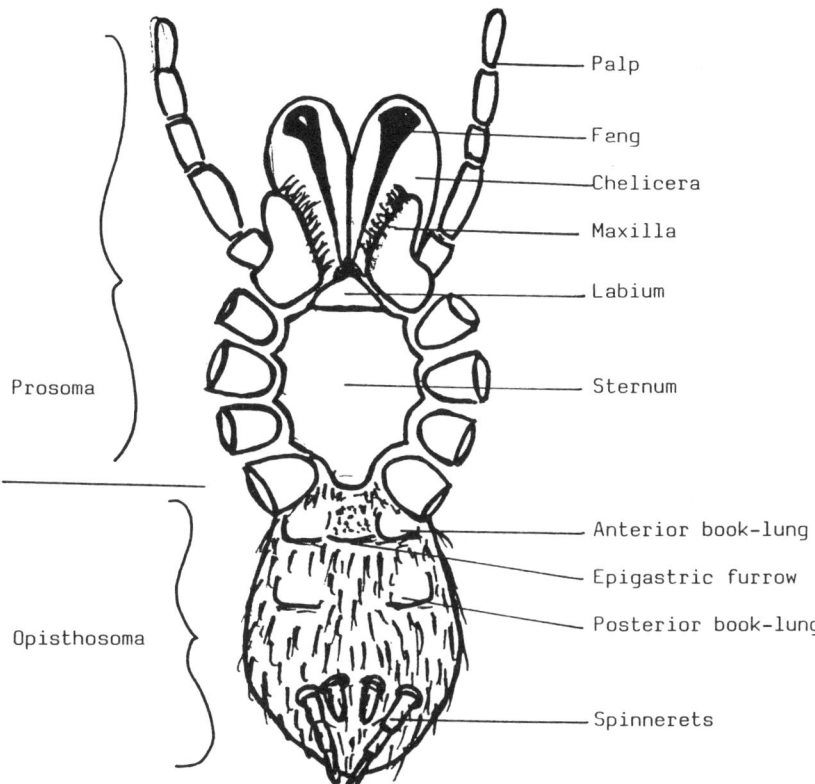

Fig. 4 Ventral view of a Theraphosid spider.

are armed with teeth which are used to mash the prey. The poison glands are located at the base of the chelicerae and when the tarantula strikes its prey, poison is injected through the fang. The chelicerae are not only used to subdue prey and as a means of defense, but as a tool, particularly in the trap door spiders (Ctenizidae) and tarantulas which use them to dig burrows and dispose of surplus material. Female tarantulas also carry and turn their egg-sacs using their chelicerae.

The pedipalps, although very similar to the legs, are not used for walking but for the manipulation of prey. In the mature male spider they are modified as organs of copulation. The base of the pedipalps form part of the mouth parts with the inner sides being fringed with hairs that act as a filter while the spider is sucking liquified food.

The prosoma contains the central nervous system, part of the intestinal tract, a pair of poison glands and the sucking stomach as well as muscles.

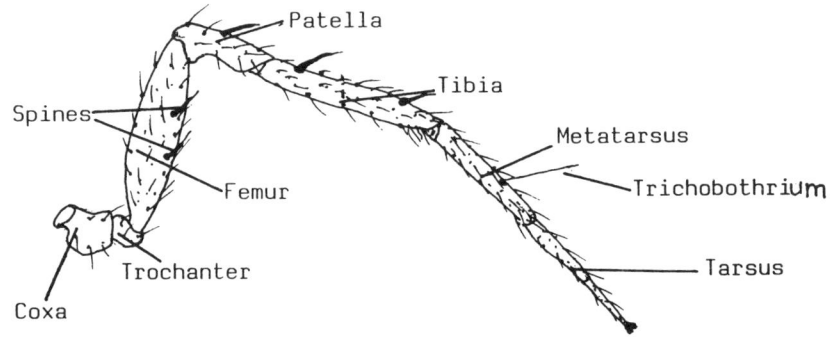

Fig. 5 Leg segments.

The point of attachment of the main muscle system can be seen on the carapace. It is the fovea which appears as dip towards the rear, or as in *Ceratogyrus darlingi*, a protrusion, hence the common name East African Horned Baboon.

The opisthosoma, or abdomen, is connected to the prosoma by a narrow pedicel. This narrow, delicate tube carries the intestines, the main blood aorta and the abdominal nerves as well as finding enough room for the returning blood. The opisthosoma of a spider is soft and expandable to allow for the storage of food and the development of eggs in the ovaries. The abdomen of a mature male is characteristically small compared to that of a female of the same species. It shows no trace of segmentation except in Mesothele, an ancient form from which present day spiders are derived. Here twelve segments can be distinguished.

On the underside of the abdomen can be seen two pairs of book-lungs, so called because they are made up of a series of 'leaves' as in a book. (Araneomorphs or true spiders have only one pair.) Oxygen is diffused through these into the blood stream. The book-lungs can be seen clearly on the inside of the shed skin of the abdomen where they appear as white patches. Between the first pair of book-lungs is a slightly raised area and a furrow or fold-like crease which runs just below these structures. This is the epigastric furrow, the site of the opening to the reproductive organs. In some species this is much more pronounced in the female than in the male. The reproductive organs will be dealt with more fully in 'Sex Determination'.

At the posterior end of the opisthosoma are the spinnerets. In tarantulas there are four, two of which are visible from above. The Dipluridae which are commonly referred to as Funnel web spiders have relatively long spinnerets as they construct more elaborate webs. *Macrothele calpeiana*, the Spanish Funnel web and *Atrax robustus*, the

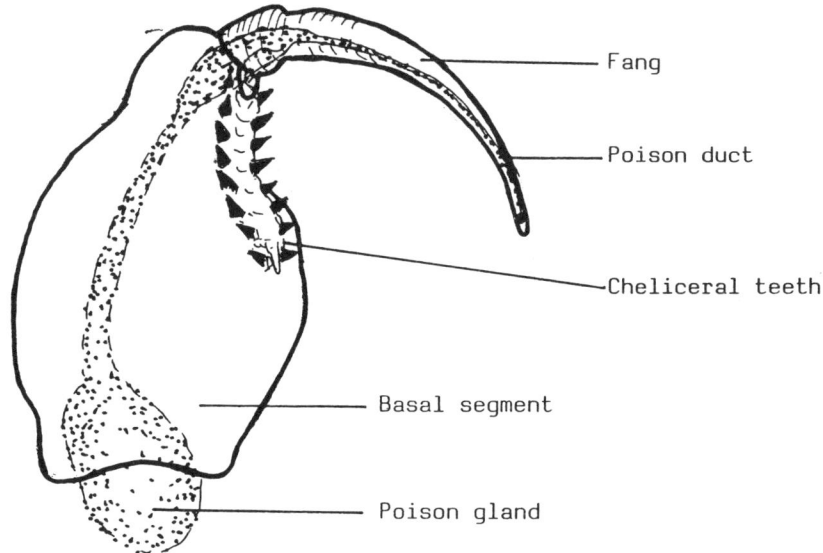

Fig. 6 Chelicera showing fang and poison gland.

Sydney Funnel web are Dipluridae. As well as producing silk to line a retreat or to make a tube web, tarantulas also produce web to make the egg-sac or sperm web.

The abdomen of a tarantula is very delicate. The skin is under pressure from the fluids within and repairing damaged skin is virtually impossible. If such damage is caused, for example by the tarantula falling or being dropped, the fluids will seep out and the spider will die slowly. Attempts have been made to stem the flow by covering the wound with talcum powder, but this is rarely successful. The developing skin underneath will also have been damaged. There is no clotting agent in a spider's blood so even if a leg is damaged the vital fluids will seep out. The exception to this is when the tarantula deliberately throws off a leg. This is done between the coxa and trochanter and the wound is sealed with the aid of muscles and fluid pressure.

The venom potential of tarantulas has been questioned often. Little work has been done on the subject, probably because they have not posed a threat to humans. Like any animal bite or sting, that of a tarantula can have little or no effect on one person whilst another person is allergic to it and can be very ill. Reported bites from tarantulas have resembled a wasp or bee sting with no other ill-effects. Some of the true spiders are considerably more dangerous. A tarantula bite may not be pleasant, but it is not considered harmful. The answer is not to get into a position where one can be bitten.

**Classification** This is a very complicated subject which is still in the process of being sorted out as far as tarantulas are concerned. It has been greatly simplified here.

As there are around 650 known species of Theraphosid spider (Tarantula) an exact method of classification is essential. A species can therefore be identified accurately. Common names are not specific enough particularly if breeding is proposed. Retailers of wild caught tarantulas often invent common names which will enhance sales and some species have been sold under a variety of names. *Euathlus vagans* is known as the Red Rump or as Black Velvet. There are three species which have been sold as Red Knee or Red Leg - *Euathlus smithi*, *Euathlus mesomelas* and *Euathlus emilia*. Also, Thai Tiger is a much more attractive name than Lesser Thailand Black. Both these names refer to *Haplopelma albostriatus*.

Scientific names based on the two name system should leave no doubt as to the exact species once it has been identified correctly. The first of the two names refers to the genus to which it belongs, the second refers to species. Many species can belong to one genus. Members of a genus have certain characteristics in common, but each species within the genus will have characteristics which distinguish it from other species. In scientific literature the generic name and specific name are followed by a name and date in brackets. This is the author who first described the species in a scientific paper and the year in which it was published.

As an example, take one of the best known tarantulas *Euathlus smithi* (Cambridge 1897), the Mexican Red Knee. Euathlus is the genus to which it belongs, along with other species having certain characteristics in common. Smithi is the species name and all tarantulas with this specific name will be of the same 'breed'. The person who first described the species in a scientific paper was O.P. Cambridge and this was published in 1897. Copies of the original papers and type specimens are held in museums of natural history throughout the world, often in the country of origin of the species, but the British Museum of Natural History has the largest collection in the world. The specific name often relates to the finder of the species. *Euathlus smithi* was discovered by a Mr. Smith. It can also relate to a distinguishing feature of the species, for example, *Haplopelma albostriatus*, meaning white stipes.

The classification of tarantulas is in a bit of turmoil at present as the whole system is being ammended. Much of the original work was done in the late 19th and early 20th centuries. The specific names do not change, but sometimes the genus is altered. Classification is based on evolutionary features. (Fig. 7) In the Order ARANEAE there are three Sub-orders. Mesothelae are very primitive spiders still showing evidence of segmentation. Orthognatha (Mygalomorphae) or tarantulas have downward striking fangs and two pairs of book-lungs. Labidognatha (Araneomorphae), the true spiders, is the most highly developed Sub-order having grasping fangs one pair of book-lungs and trachea. The Sub-order Orthognatha is further divided into eleven Families which include Theraphosidae (tarantulas), Ctenizidae (trap door spiders), Dipluridae (funnel web spiders) and Atypidae (purse web spiders). The Family Theraphosidae is further divided into seven Sub-families. These are Aviculariinae, Eumenophorinae, Grammostolinae, Ischnocolinae, Ornithoctoniae, Theraphosinae and Selenocosmiinae.

This is how the example *Euathlus smithi* fits into the order of things.
Class      – Arachnida
Order      – Araneae
Sub-order  – Mygalomorphae

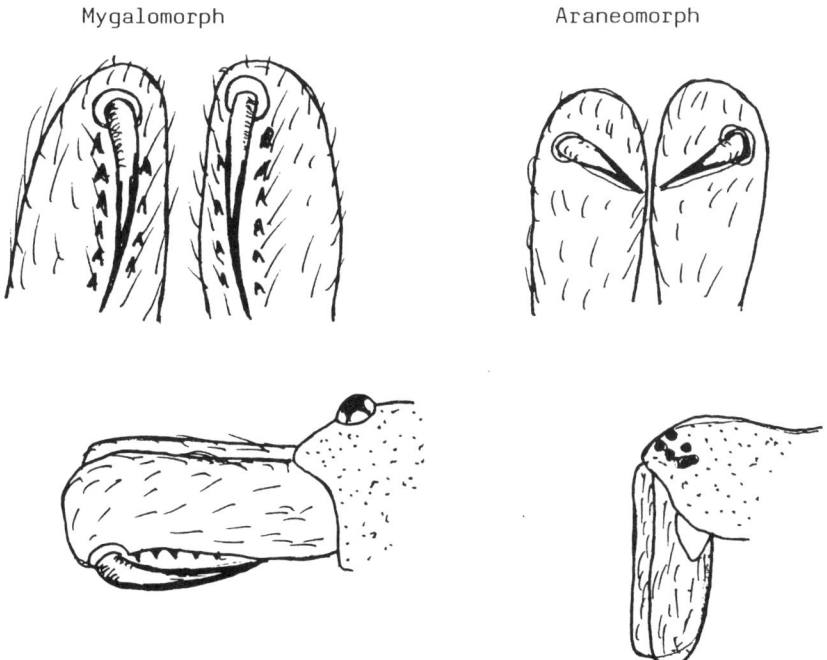

Fig. 7 Fangs of Mygalomorph and Araneomorph spiders.

Family       – Theraphosidae
Sub-family – Grammostolinae
Genus        – Euathlus
Species      – smithi

Classification is based on evolution and the stage of development a particular species has reached. However, apart from a basic understanding of where tarantulas fit into the overall scheme, it is a subject in which the amateur can become very confused.

**Identification** Identification of known species is much more relevant than classification to the hobbyist. This is based on measurements and distinguishing features. The difference between two closely related species may only be the length or position of some hairs, but proper identification is vital for successful breeding. The methods used are:-
 1. Metatarsal and tarsal scopulae.
 2. The arrangement of the eyes.
 3. Leg segment and carapace measurement.
 4. Foveal groove on the carapace.
 5. Stridulating organ.
 6. Secondary stridulating organ.
 7. Colour.
 8. Genital organs.

*Metatarsal and tarsal scopulae* As seen earlier, these scopulae or hairs enable the tarantula to walk on smooth vertical surfaces. The number, arrangement and length of these hairs vary between species. They can be measured using calipers and a millimetre ruler.

*Ocular arrangement* The position of the eyes on the carapace varies from species to species. They are arranged in roughly two lines, anterior and posterior. Those on the outside are referred to as lateral and those on the inside as median. (Fig. 8)

*Leg segment and carapace measurement* The leg segments and carapace are measured and considered in relation to each other. A small specimen will be of the same relative proportions as a large specimen of the same species. The leg segments are measured as follows:- femur, patella, tibia, metatarsus, tarsus. Patella and tibia are added, as are metatarsus and tarsus. These measurements are used to compare with those of the carapace.

*Foveal groove* This will, in the majority of species, fit into one of three types, transverse, crescentic or procurved. (Fig. 9) The exceptions are

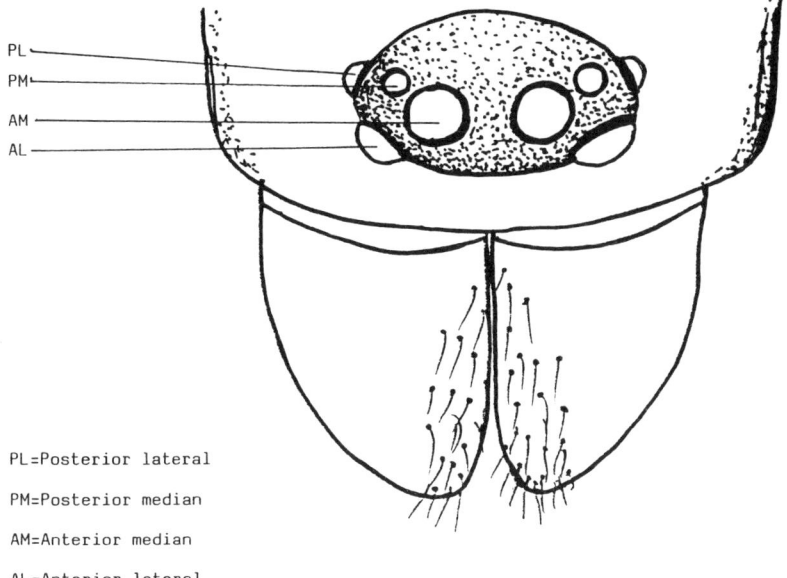

Fig. 8 Ocular arrangement.

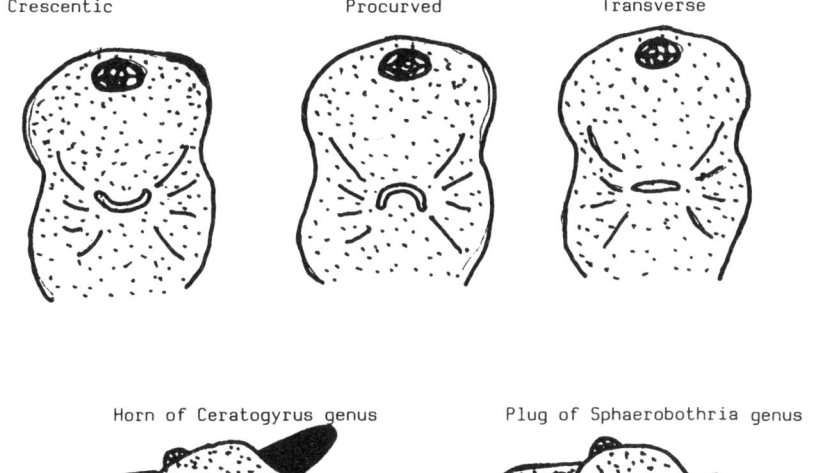

Fig. 9 Foveal groove.

Ceratogyrus species and Sphaerobothria species. These do not have a groove, but have a horn or plug.

*Stridulating organ* The stridulating organ is the means by which some species can make a hissing sound. This is done by rubbing the chelicerea and maxillae together so that the bristles or hairs vibrate. The hairs at the base of the chelicerae and the maxillae are examined.

*Secondary stridulating organ* These hairs are situated between the base of the pedipalp and the first leg. They are similar to those on the chelicerae.

*Colour* This is not a reliable means of identification as not only do individuals vary within a species but the colour of a spider varies according to the date of its last moult. Colour can only be used as a rough guide.

*Genital organs* Examination of the sexual organs of Araneomorphs is the accepted method of identification as each species is unique to prevent cross mating. The genital organs of Theraphosidae are not so highly evolved and elaborate and there has not been a great deal of work done on this aspect. The moult skin of a mature female will reveal the spermathecae, each species having a unique shape. (See 'Sex Determination). Otherwise the epigyne of a dead specimen can be cut out and examined under a microscope. The male is more difficult because he only has mature palps for the period following his last moult. The palp is unique to the species and can only fit into the epigyne of a female of the same species. The male must be dead and the palp removed. The embolus is extended using lactic acid and it can then be examined under a microscope.

The amateur arachnologist needs to use a method of identification that does not require the specimen to be dead, particularly if he wishes to breed. Unfortunately the most positive and reliable method is to dissect a preserved specimen and to use a microscope. However, species are often imported in batches of quite large numbers from a particular area. Inevitably some die and if these are used for identification it is 99% positive that the rest of that batch are of the same species. With the availability of captive bred tarantulas, identification has been carried out on previous generations.

Almost as good as a dead specimen is the moult skin of a living spider. Although this is fragile and must be handled carefully all the external features are in tact, as well as the internal sexual organs of the epigyne. In fact, of all the methods described, only the male palp cannot be examined. A living spider can be examined closely if it is placed in a plastic bag with most of the air removed. If it is handled carefully it should come to no harm, but in this case the genital organs of neither sex can be examined.

As many of the methods as possible should be used on each specimen to reduce the chance of error. Identification is not an easy exercise for the inexperienced and a lot of practice is needed. Many of the scientific papers which were written in the last century are inadequate and need to be revised. A large number were written in Spanish and Portugese and have lost accuracy through translation. As more information becomes available the task should become easier.

Chapter Three

# Choosing a Tarantula

The reasons for wishing to keep a tarantula should be examined and careful preparations made before one is purchased. It is only through doing this that problems will be minimised.

A tarantula must always be considered as a wild animal and not as a cuddly pet. Handling is not recommended. Neither should a tarantula be used by its owner as a means of frightening other people. Spiders are feared by many for a variety of reasons. It may be their predatory habits, their ability to move fast or the number of legs. Although this fear may appear to be illogical to the spider enthusiast, it is very real to those who suffer from arachnophobia. The media play their part in perpetuating this fear where spiders have been used to induce terror. James Bond apparently found himself in a life-threatening position when a particularly docile species of tarantula was put into his bed. The film 'Arachnophobia' released recently, although quite entertaining, did not do much to help those with a fear of spiders. A little knowledge of the behaviour of spiders resulted in it being rather amusing. Portrayals of spiders in such a way reinforces people's fear of them and irresponsible owners may use their pet to frighten others. A tarantula may also be acquired as a status symbol, an interesting ornament. This ornament is a living creature and must be cared for correctly.

As with the acquisition of any pet, there must be long term considerations. With some tarantulas living for twenty or more years there must be a commitment for this length of time. Buying a tarantula on the spur of the moment is not a good idea. Preparations should be made and living quarters got ready before the spider is bought. One would not buy tropical fish without an aquarium or a budgie without a cage.

Once the decision has been made to keep a tarantula, living quarters must be prepared and monitored beforehand. (See 'Housing'). A source of live food must be located. Most important is a knowledge of their habits and requirements.

Not all species of tarantula are suitable for the beginner. Some are quite delicate and intolerant of less than optimum conditions. Others can be very fast and rather aggressive. More detailed information on individual species is given in the 'Species File', but a good beginners tarantula is *Grammostola spatulatus*, the Chilean Rose. These are usually easily obtainable from pet shops and are inexpensive. African Baboon spiders and Asian species should be avoided by the beginner. Unfortunately some pet shop owners do not know what species they are selling – they are all tarantulas, so try to find a reputable dealer.

Many traders are becoming more enlightened as to the needs of tarantulas and a reputable dealer will be pleased to give advice, as well as keeping healthy stock. However, with wild caught stock, it is impossible to know the past history of the animal. There is no way of telling its age and therefore know what its life expectancy will be. The only exception is in the case of a mature male which will be complete with mating hooks and fully developed palps. Unless required for breeding, the mature male is of little use to the pet keeper as it will only have a short time to live.

Sex is an important factor when choosing a spider as the males only live a short time after maturing whereas a female may have many years ahead of her. Unfortunately it is impossible for the amateur to determine the sex of an immature tarantula. What appears to be a plump female may quickly mature into a male. The only positive way of knowing that a tarantula is a mature female is if she is carrying an egg-sac. Otherwise the only reliable method is to examine a shed skin. A reputable breeder can often supply a specific sex.

It can be a gamble purchasing an imported tarantula. They could have spent a considerable time after capture in cramped conditions without food. They have been subjected to the trauma of being transported half way round the world. Wild tarantulas can be carrying disease (not transmittable to humans) or parasites. These parasites may not be apparent, but sometimes take the form of a lump on the spider's abdomen. Inside this lump could be the larva of a parasitic wasp, the wild tarantula's greatest enemy. This larva will be feeding on the spider and when the spider is most vulnerable, for example when moulting, will emerge in its adult form. The spider will die. Tarantulas with lumps on their abdomen should be avoided.

A tarantula, even of the most docile species, should be alert. If it has a hunched appearance and is sluggish in its movements it could be unwell having been kept too long in poor conditions and at low temperatures. Many of these unfortunate creatures do not survive. However a tarantula which may appear to be old may only be coming up to a moult. At this stage they look dull and drab and often have a bald patch on the abdomen through throwing hairs. It is worth taking a chance with such a spider because when it does eventually moult it will be perfect, bright and the abdomen covered with hairs again. The same applies to tarantulas with missing limbs. A seven-legged spider may be less expensive than a complete one, but on moulting the missing limb will be regenerated.

To avoid the possibility of buying an aged tarantula, it is better, if possible, to buy a captive bred one. The age should be known, as well as the sex, and the spider will be free of parasites and disease. Also, a captive bred spider has never experienced the freedom of the wild and will be used to captivity therefore more content. Sub-adults are best for the novice. They have survived all the difficulties of the spiderling stage and still have their whole adult life ahead of them. They will also be large enough to be housed in adult quarters.

When the tarantula has been brought home, particularly if it is a wild caught one, it should be given the opportunity to drink as it may have become dehydrated. It can be placed in a container with a shallow drinking dish. If it is thirsty it will sit over this dish and drink. Before the tarantula is placed in its living quarters there is an opportunity to examine it more closely. It can be placed in a clear plastic container or in a plastic bag. All parts can then be seen, including the underneath, without actually touching it. A docile species can be encouraged to walk into the bag or container.

When the tarantula is eventually introduced to its permanent housing it may react in one of two ways. It may go into a corner and 'freeze' for a few hours before taking its first tentative steps on unfamiliar ground, or it may immediately try to escape. The lid must be to hand in case of the latter happening. When it begins to explore its new home it will begin to lay web. In some species this carpet is barely noticeable, but it is similar to the spider lining its burrow or retreat. Other species will begin to form their tube retreat from web. It is only when the spider appears settled, having had a day or so to establish its territory, that food should be introduced.

Handling is not recommended, the main reason being that it is dangerous for the tarantula. If it falls its abdomen will burst and the animal will die, probably slowly. Legs caught in clothing may be thrown off and, although these will be regenerated, it is unnecessary damage to the spider. Handling can also frighten the spider so it will throw hairs from its abdomen resulting in a bald patch for it and a nasty rash for the handler. Once the tarantula is settled and has proved to be docile, it can be picked up by encouraging it to walk onto a hand. Great care must be taken as they are unpredictable and may try to escape. The safest place to change the tarantula from one container to another is in the bath as it will be unable to run up the sides. It can be gently encouraged to walk into the new container so handling can be kept to a minimum.

A tarantula is an ideal pet for animal lovers who, for one reason or another, cannot keep a more conventional pet. They are ideal for flats as they do not take up much space. They do not make a noise to disturb the neighbours, and do not need to be exercised. Where dogs and cats are prohibited a tarantula can often be kept. If only one member of a family has a liking for spiders, a tarantula can be kept in a bedroom where it will not frighten the others. A tarantula can be left alone for a few days without coming to any harm, for example during the owner's holiday. An escaped tarantula does not pose a threat to anyone as it will not survive outside the controlled environment of its housing.

The reasons for wishing to keep a tarantula should be examined closely, just as for any pet. Their requirements should be understood before acquiring it. Every effort should be made to obtain the tarantula from a reputable source. If properly cared for, these large spiders can give many years of interest and pleasure and many people who start with one become so interested that they expand their collection and breed them.

Chapter Four

# Housing

When considering the type of housing suitable for a tarantula an understanding must be had of the natural environment in which each species lives. It is important that these natural conditions are reproduced as closely as possible for the spider to be kept healthy and happy. Species can be broadly divided into three categories – burrowing, arboreal and opportunist. Burrowing species may spend their entire life in the wild in one burrow, enlarging it as they grow. Arboreal species, as the name suggests, live in trees, making their tube webs behind bark or in leaves such as banana. Opportunists do not make a burrow, merely a 'scrape' under a rock or log, in an old lizard burrow or even under wooden floors of houses. (Figs. 10, 11, 12)

No tarantula, even those originating from arid regions, can withstand very high temperatures or a dry atmosphere. The outdated idea of the spider in an aquarium, sitting on sand with a light bulb blazing overhead and a cactus plant in the corner is nothing short of cruelty. In their natural environment such species will live deep in a burrow. They will only emerge when the heat of the day has passed. The burrow provides an even temperature with little daily variation at its deepest. Also, moisture is held in the ground at this depth and this provides the humidity necessary for the spider to be able to breathe. In these natural conditions the tarantula can very effectively regulate the temperature by moving up or down the burrow. It would never emerge when the sun was at its height unless to escape a predator.

The cactus plant can be a potential killer. Tarantulas are capable of climbing glass and if they fall can impale themselves on the spines of the cactus. Plants in general are best avoided as they can lead to problems arising due to the high humidity, particularly mould. The tarantula will not be interested in decor, but from the owners point of view the container

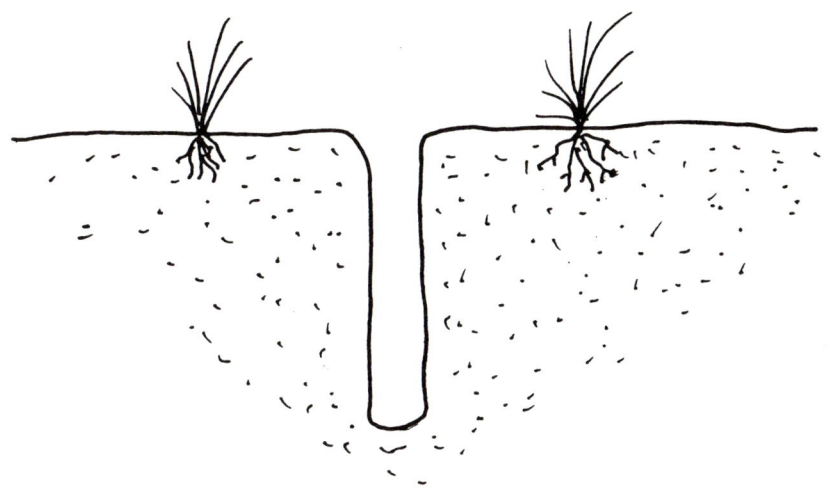

Fig. 10 A burrow of a Theraphosid spider.

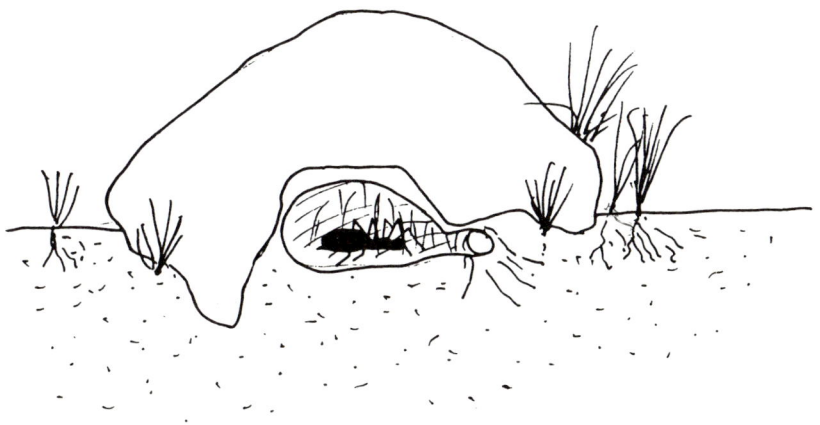

Fig. 11 A 'scrape' under a rock.

can be enhanced by plastic plants or cork bark. These lend themselves more readily to containers for arboreal species. Even though certain species will spend most of their lives in a single burrow, they appear to regard their container as the burrow when in captivity. The majority of adult specimens do not attempt to construct a burrow, but will use an artificial retreat such as a flower pot or a piece of bark. It is difficult to provide a substrate identical to that in the wild and, with periodic cleaning the burrow would be destroyed. It is also thought that some

Fig. 12 An arboreal web.

species begin a burrow as spiderlings and enlarge it as they grow, making use of a worm hole, for example. Others may lose the ability to construct a burrow as they become older because in the wild this would not be necessary. This, of course, can be advantageous for the owner, because the pet in a burrow would rarely be seen.

The pet tarantula is in an artificial environment and it is very difficult to reproduce the natural conditions exactly. Because of this great care should be taken in the preparation of the spider's living quarters bearing in mind temperature, humidity and space available.

**The container** The living quarters should provide as near natural conditions as possible. Of secondary importance is the appearance. What may be considered an attractive addition to the owner's living room may be totally unsuitable for the tarantula.

The first consideration must be size in relation to the species and age of the tarantula. The average adult does not require a large container as, in the wild, they do not wander far. If housed in a large container the spider will occupy a corner and not stray far from it. The ideal size for the average adult (e.g. *Grammostola spatulatus*, the Chilean Rose or *Euathlus smithi*, the Mexican Red Knee) is 25 cm. × 25 cm. × 25cm. The spider will have adequate space without feeling vulnerable. It will also have a greater chance of catching its food as the majority of species are quite lazy. Horror has been expressed by people who do not know anything about tarantulas concerning the lack of space they are given. In fact it is cruel to give them much more than the recommended size.

There is a wide range of glass aquaria available in pet shops. These can be purchased ready-made or can often be made to order to the required dimensions. The size quoted is not mandatory, but variations on a similar volume can be made. Containers can be made at home using glass and aquarium sealant. This is particularly useful if required to fit into an awkward space. An essential part of the container is a close fitting lid. Tarantulas can escape through very small spaces and they are also strong enough to lift an inadequate lid. Climbing glass is no problem. The lid should allow easy access for feeding and watering.

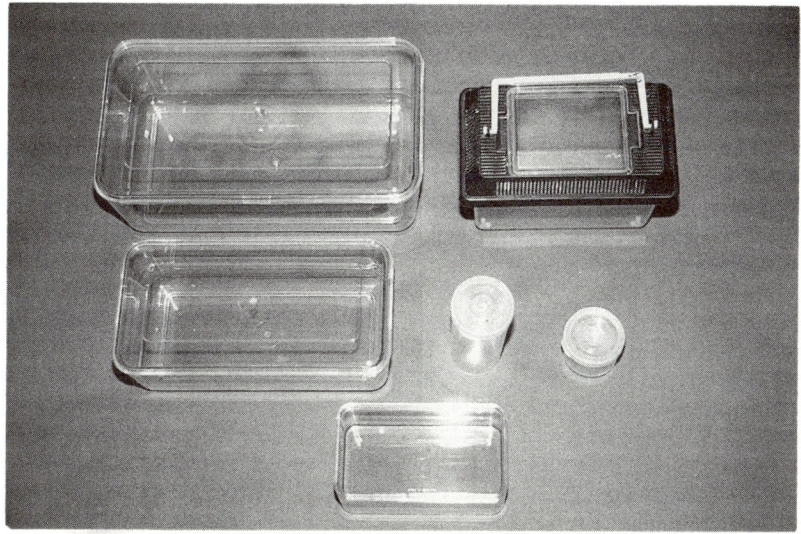

Fig. 13 A range of plastic containers suitable for spiderlings to adults.

Plastic containers are also widely available. These may be purpose made for small pets or the owner can improvise by using suitably sized clear plastic food containers. The pet containers have a secure lid, but this often allows too much ventilation so it is difficult to control the humidity. To overcome this plastic sheeting can be cut to size and stuck to the underside of the lid. An adequate amount on air will still be able to circulate. Food containers have no ventilation and the lids are not very secure. Small holes can be drilled in the plastic sides using a 2mm. drill bit. The lid can be held down with tape. Plastic containers will provide a much greater range in size and shape than ready made glass aquaria and can be easily stacked if a number of tarantulas are kept. Plastic has an advantage over glass in that it is more suitable for attaching web and is a warmer material. Its disadvantage, however, is that it can, in time, become scratched. When a number of spiders are kept, a large aquarium can be divided using glass and sealant or a sheet of rigid plastic. It must be impossible for the tarantulas to move from one section to the other – one will eat the other! This method economises on space and makes heating simpler.

Although tarantulas are highly predatory creatures, some species, given enough space and food can be kept together. *Avicularia avicularia*, The Pink Toed tarantula and *Grammostola spatulatus*, Chilean Rose have been kept successfully together. (See 'Species File').

One of the problems with conventional style housing is that often the tarantula cannot be seen, particularly if it is inclined to burrow. As the amount of space required surrounding the burrow is minimal and depth is more important, artificial burrows can be constructed. An ideal material for this is found in Durox building blocks. These are relatively light, are easy to saw and can be carved into the desired shape. The ground in which the wild spider will burrow is usually very hard so these blocks will be very similar to the natural environment. They also hold humidity. The following is an example of one constructed for a *Haplopelma minax* (Thailand Black) in which the spider lived for a number of years. First a wooden frame was made out of 5cm. $\times$ 1½cm. wood. It was 65cm. high by 18cm. wide by 5cm. deep. A plywood back was attached. A piece of block was cut measuring 35cm. $\times$ 15cm. $\times$ 5cm. The block was then carved to form the burrow and thoroughly washed to remove any dust. It had absorbed enough water to produce an acceptable humidity level. It was then placed in the frame with 'ground level' 15cm. below the top and the space remaining below filled with sand. A feeding/watering hole was drilled in the top of the frame and a plug fitted. Ventilation holes were drilled in the sides above the block and these were covered with fine wire mesh. Natural materials (sphagnum moss, branches) were added at the top. A glass front was attached with brackets so that it could be easily removed for cleaning. The whole thing can be hung on a wall. (Fig. 17)

By constructing housing with the spider in mind, a very natural environment is reproduced and the tarantula will live happily, mostly in a vertical position, and web over the walls of the burrow, just as it would do in the wild. It will come to the surface to feed at night, but can still be seen during the day. If the spider is initially reluctant to use the burrow it can be encouraged to do so by covering the glass with black paper for a few days. The whole set-up can be cleaned regularly by dismantling and washing thoroughly while the spider is held in a small container for the short time it takes. Humidity can be maintained by wetting the block, and food introduced through the feeding hole. A similar set-up can be used to house scorpions. Another versatile material for providing artificial burrows is that used by florists to make floral arrangements. It is not as heavy as the building blocks, but is only available in relatively small sizes. It can be placed in a conventional container with the burrow carved so that one side is open and this is positioned against the side of the container. Because this material is light a determined tarantula will have no difficulty in moving it if it is smaller than the container.

If spiderlings are being reared, small plastic tubs 5cm. tall by 5cm. in diameter are ideal. When these are half filled with substrate a small spiderling can feed easily. A large number of these tubs can be placed in a larger container to facilitate heating. As the spiderlings grow they can be moved to slightly larger containers in stages.

Arboreal species of tarantula require height rather than floor area as well as twigs and bark on which they construct their web. If the bark is placed close to the side of the container and that side temporarily covered with black paper the tarantula will be encouraged to make its web there. Given a couple of weeks it will be settled and the paper can be removed. Immature specimens can be kept in containers such as plastic sweet jars. If the jar is set on its lid cleaning will be much easier as the substrate can be removed without disturbing the spider. Ventilation holes can be made around the top as well as a feeding hole. A piece of bark can be screwed to the side of the jar. Spiderlings can be housed in plastic tubes 5 cm. in diameter and 10 cm. tall with a twig or piece of plastic sheeting to which they can attach their web.

The major consideration when selecting a container for housing must be hygiene. The material must be easily cleaned. For this reason glass and plastic are more suitable than natural materials such as wood. If wood is used it will eventually rot in the high humidity, which is unfortunate as attractive containers can be made using wood with a glass front. This problem does not arise when combining wood with building blocks as the block does not release so much water into the wood as another loose substrate such as Vermiculite.

**Heating** Most commonly kept species of tarantula originate from tropical latitudes, particularly the larger ones. Heat must be provided if the ambient temperature is less than 20°C (70°F). The optimum temperature for tropical tarantulas is between 23°C and 26°C (75°F to 78°F) with a slight drop at night. They will survive for short periods with drops to 15°C (60°F) but this is not recommended as spiders cannot regulate their own body temperature and when they are cold they are sluggish and will not feed. On the other hand, excessive heat will kill them more quickly. This tends to increase the rate of water loss and also the rigidity of the exoskeleton prevents the expansion of the internal fluids with the result that the vessels and ducts are closed under pressure. The younger the spider, the more quickly it is effected by a rise in temperature.

Whatever method of heating is used a reliable thermometer and thermostat are vital. In the wild the tarantula will be able to regulate the temperature to a certain extent by moving up or down its burrow. The captive tarantula has no means of doing this as it is unlikely that the substrate will be deep enough. If a heating pad is used, it will have nowhere to go. It is very important that the heating is correctly regulated before the spider is introduced.

Various methods of heating can be used, the most effective and natural being to heat an entire room. For the enthusiast who keeps a large number of tarantulas this is also the most cost effective means of heating. If the room is well insulated costs should not be too high. It is important when considering a heat source that it also allows for movement of air around the room for a more even temperature. A thermostatically controlled fan heater serves the purpose well and some models can be used as fan only during spells of hot weather. The setting should be checked regularly against a thermometer. There will be some areas within the room where the temperature is lower than others, for example against an outside wall. With careful arrangement of species, according to the temperature of their natural habitats all available space can be utilised. This method of indirect heating may be ideal for a number of tarantulas, but is impractical if only one or two are kept. An adult of the more hardy species such as *Euathlus smithi* (Mexican Red Leg) or *Grammostola spatulata* (Chilean Rose) will live quite happily if the container is placed in a warm part of the room, for example beside a radiator, as long as there is not too great a fall in temperature at night. Direct sunlight should be avoided as the container, particularly if glass, will act as a greenhouse and the spider will overheat. As with all methods of heating, temperature should be monitored BEFORE the spider is introduced and a careful check kept to ensure it is maintained around 23°C (75°F).

When the number of tarantulas being kept exceeds that where it is practical to heat each one individually, a heated cabinet can be used.

This can be constructed quite cheaply, or a cupboard can be converted. Running costs can be kept to a minimum if it is well insulated. Heating can be provided by a heating cable or by a light bulb which does not shine directly on the spiders. If the cabinet is fitted with glass doors and a fluorescent light on a time switch, it can look quite attractive. So that the containers are not sitting directly on the heat source, if pads or cables are used, they can be raised on blocks of wood. This will allow the heat to circulate more freely. Once the temperature has been monitored at different levels species can be arranged accordingly. This method is particularly suitable for rearing spiderlings or juveniles which are less tolerant than adults of temperature variations and it is impractical to heat each container. As always a reliable thermostat is essential.

Direct heating can be provided in a number of ways, either from above, below or a side. To reproduce natural conditions as accurately as possible the heat source should be from above, representing the sun, and this can be done by using a light bulb. Certain precautions must be taken, the most important being to monitor the temperature before putting the spider in. It is essential that a thermostat is wired in and set so that it will turn the bulb off when the temperature reaches 26°C (78°F) and on again when it drops below 22°C (73°F). A fairly low wattage bulb should be used, no more than 40W to maintain a steadier temperature and the bulb should be coloured, red being best. Tarantulas do not like bright lights and will be more inclined to venture out of their retreat in the dark or in subdued lighting. This method of heating does not allow for 'night' and 'day', most tarantulas being used to a twelve hour cycle. If a false bottom is put in the lid, made of an opaque material which will not become hot, the light will not interfere with the daily cycle. It also overcomes the problem of having a naked light bulb on which the spider can burn itself. The arboreal species in particular, such as *Psalmopoeus cambridgei*, the Trinidad Chevron, have no problem in walking across the roof of the container and coming into contact with the bulb. Alternatively the bulb can be protected with netting. (Fig. 14)

The other method of providing direct heat is by means of heating pads placed underneath the container. Again these should be wired up to a reliable thermostat. Pads have the distinct disadvantage of having the heat source below the spider which is very unnatural. As ground dwelling species will burrow to cooler levels in the wild if they become too hot, using heating pads will make the lower levels hotter. To reduce this problem the pad could be half the area of the container, leaving one side of the substrate cooler, but the air would still be heated. A fairly deep substrate should be used when heating from below. If the container is raised about $\frac{1}{2}$cm. above the pad it will avoid damaging the heating

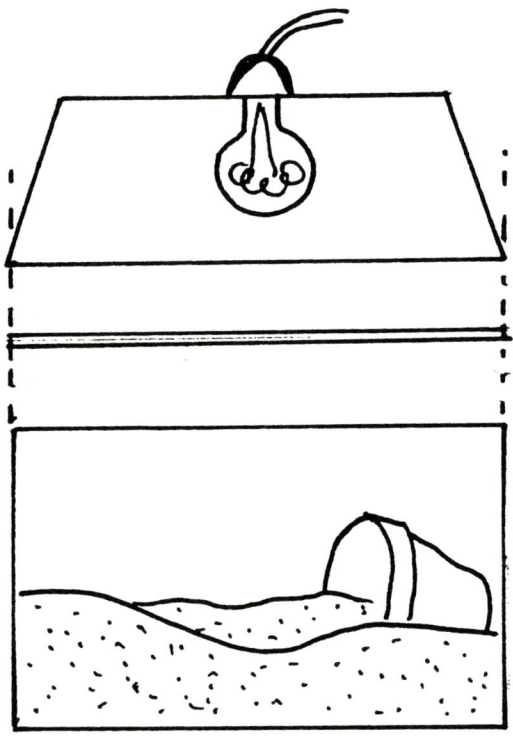

Fig. 14 Light bulb heating.

element, or the risk of cracking the bottom if a glass aquarium is used. (Fig. 15)

Heating cables, similar to those used in the propogation of seeds, give a less concentrated heat source and can be arranged so that part of the substrate is unheated. These are particularly useful if a number of containers require heating. These can be placed on a tray with the cable running under them. Only one thermostat will be needed, in one container, as they will all be heated to the same level. Both cables and pads are more suitable for arboreal species which have no inclination to burrow. (Fig. 16)

A hot rock can be used if the container is large enough. These are available from some pet shops and if the container is fairly well sealed the temperature will be maintained. It should never be allowed to become so hot that it is a danger to the spider and it is advisable to have it covered so that the spider cannot walk on it.

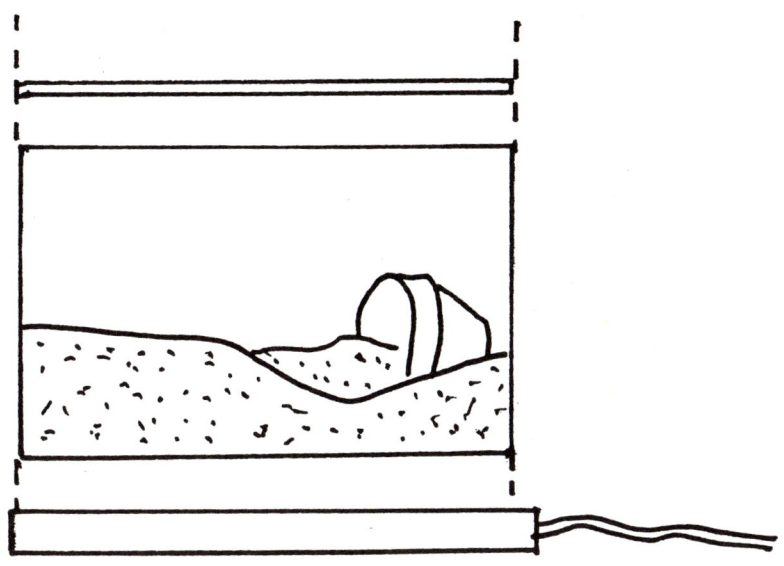

Fig. 15 A heating pad.

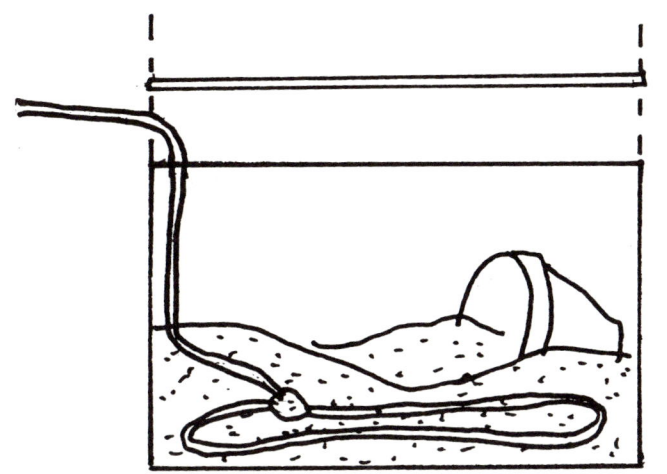

Fig. 16 A heating cable.

Submersible aquarium heaters can be used. The heater is placed in a jar of water which is sealed except for space for the electric cable. The temperature is adjusted using a thermostat and as the water gets hotter it heats the air in the container. A more elaborate way of using this method is to put a false back in the aquarium using a sheet of glass and sealant making the space about 4cm. wide. This space is then filled with water and the heater placed in it. Problems which could arise with hot rocks and submersible heaters are that if only a small volume (rock or water) is used to heat a large space, the 'heater' may become dangerously hot before the required air temperature is reached. Care must be taken to prevent the spider touching the heat source.

Greater care must be taken to maintain the correct temperature for spiderlings as they are less tolerant, particularly of high temperatures. An aquarium with the individual containers placed inside can be heated by bulb, pad or cable. It is important to remember that those spiderlings nearest the heat source will be hottest, so careful temperature readings should be taken in different parts of the aquarium.

Whatever form of heating is used it is vital that a reliable thermostat is connected to it and a thermometer is used. This can either be a conventional mercury or alcohol type or a digital one which can be attached to the side of the container. Before a tarantula is purchased the container should be set up and tested. Only when everything is functioning correctly should the spider be put in.

**Humidity and watering** No tarantula will survive for long in a completely dry atmosphere. Apart from becoming dehydrated, its respiratory system will be unable to function. A tarantula, unlike the more advanced true spiders, has no trachea or internal organs for carrying oxygen around its body. It has two pairs of book-lungs which diffuse the oxygen into the open blood system. These book-lungs, working on a similar principle to fishs' gills, must be kept moist in order that the oxygen can pass into the blood. They will not function in conditions of very low humidity and the spider will 'suffocate'. Hence the tarantulas of arid regions live in deep burrows. Humidity requirements vary between species. *Theraphosa leblondi*, The Goliath Bird Eater, originates from swampy areas in the tropical rain forest so it requires anything up to 100% relative humidity. Alternatively, *Ceratogyrus bechuanicus*, the Starburst Horned Baboon, comes from dry woodland and survives best at 70% relative humidity. More detailed information on specific species can be found in the 'Species File'.

A reliable relative humidity gauge which can be attached to the container must be used. These can be bought either as an adhesive strip or a dial and are obtainable from pet shops. The humidity can be regulated by the amount of water added to the substrate, thus the need for an absorbant material. If water is added to half the floor area for ground dwelling species, it will be graduated from damp to dry so giving the spider a choice as to the most comfortable area for it. It will need the damper area when moulting. The entire floor area can be moistened for arboreal species. If the correct humidity level is maintained there is no need to provide the tarantula with drinking water. This may appear unkind, but how many spiders in the wild are supplied with dishes of clean water? If it is felt necessary to provide drinking water it must be kept clean and changed daily. The dish should be shallow and have a diameter no larger than the spider's leg-span. It is a mistake to use damp cotton wool as the fibres can become entangled in the tarantula's delicate mouth parts. From experience, many tarantulas overturn the dish and others have been seen to deliberately use it as a toilet! Refilling an overturned dish will eventually lead to too high a level of humidity. The only time it should be necessary to give a tarantula a drink is if it has become dehydrated, perhaps in transit.

In the wild a tarantula will be able to find adequate water either from the sides of the burrow, from dew or from rainfall. In captivity when the temperature drops slightly at night, water condenses on the sides of the container. This gives the spider pure, clean water without any additives which may be found in tap water. They can also absorb water from the substrate. Arboreal species will naturally drink from dew or rain. The practice of spraying the container is popular, but this does tend to frighten the spider which may try to make its escape while the lid is open. A gentler way is to use a squeezy bottle with a fine curved spout. The water can be directed accurately onto the bark, web or sides of the container without disturbing the tarantula. The same thing can be used for topping up the humidity for other species. If the level of humidity becomes too high and the substrate too wet the spider will be uncomfortable and even die. With too much condensation forming on the sides spiderlings can become trapped in the film of water and be unable to escape from it. This problem can be overcome by adding more dry substrate to absorb the excess water. This is particularly effective if Vermiculite is used. Relative humidity is a very important factor in keeping tarantulas successfully and should be monitored carefully. Too high or too low, problems will arise.

**Substrate** The substrate is the material put into the tarantula's container to resemble as closely as possible the ground on which the wild creature lives. Unfortunately it is impossible to reproduce the natural substrate, but a variety of materials can be used. They serve a dual purpose – to allow the spider to burrow if it wishes and, most important, to regulate the relative humidity. It cannot be left in the container for the tarantula's life, so if it does burrow it will have to repeat the process when the substrate is removed and replaced. It is essential that this is done at regular intervals to prevent the possibility of disease, mites and fungi. In the wild the natural balance of the ecosystem takes care of these problems but in artificial conditions they can be disastrous, especially if a number of tarantulas are kept in close proximity. No matter how carefully food remains are removed, traces will be left and the high humidity provides ideal breeding conditions for mites. The substrate of an adult tarantula should be replaced at least twice a year. However some species, for example *Haplopelma lividus*, the Burmese Blue, require cleaning more frequently. The waste materials produced by tarantulas do not present much of a problem as they excrete chalky, odourless material, often on the side of the container.

Organic materials such as peat make an attractive natural substrate but they can lead to problems, particularly mites. A temporary solution is to sterilize the material before use, either in a microwave or in a conventional oven at high temperature. Unfortunately it will not remain sterile for long and, being organic, will attract and support mites. If used it should be changed frequently. Spiderlings can burrow in peat, but adults find it difficult as larger burrows collapse.

Vermiculite is an inorganic material obtainable from garden centres. It is used for potting plants. It is sterile and remains free of pests for much longer than peat. Unless contaiminated with decaying food, it provides no sustenance for mites. It therefore does not need to be changed so frequently. It also has the advantage of retaining moisture, although it is not suitable for adult spiders to make a burrow, being too light. Spiderlings readily burrow in it. It can be obtained in three grades – fine, medium and coarse. Fine grade is the best to use, particularly for spiderlings and juveniles.

Sand or gravel are often used as a substrate, but the latter is totally unsuitable. It does not retain water satisfactorally and it is impossible for a spider to construct a burrow if it wishes. Sand is more suitable, particularly fine pit sand, as it allows the spider to dig and it holds its shape well. Care must be taken that it does not become waterlogged at its lower levels as it will become black and foul due to lack of oxygen. Sand is heavy if the container needs to be moved for cleaning, etc.

Forest bark, although an organic material, does not encourage the development of mites to the same extent as peat. This is due to the tannin in the bark which does not effect the tarantula. It should be sterilized before use to prevent the growth of fungi. It does not lend itself readily to making burrows, but is particularly suitable for scorpions.

Coconut fibre is a relatively new material available from garden centres. It is ecologically friendly as it is based on waste material and can be used in the same way as peat. Being organic it should be sterilized before use.

A mixture of materials can be used and by experimenting with quantities a suitable substrate can be prepared for individual species requirements. Sand and forest bark or peat can be mixed to produce a medium in which a burrow can be constructed. For ground dwelling species which are not inclined to burrow Vermiculite is recommended. Although many tarantulas do not attempt to construct a burrow and are quite happy regarding their container as such, they often like to rearrange the substrate and can be seen moving it around using their palps and chelicerae. Ground dwelling species also like to carpet their living area with web as this facilitates walking and marks the limits of their territory. They can be seen to do this whilst eating. Sometimes this sheet of web is so fine it can barely be seen but, unfortunately, artificial substrates do not hold it firm. This is particularly so with Vermiculite.

As a general rule the container should contain about one third of its depth in substrate. Experience of one's own spider will show whether it is inclined to burrow and allowances in depth should be made for this. Enough material should be allowed to maintain the correct humidity level and a deeper layer will not dry out so quickly as a thin layer. Because the ground dwelling tarantula does not require a lot of 'head' room, this can be reduced by adding more substrate. Although they may never walk on it, substrate must be provided for arboreal species. It does not need to be as deep as that for other species, but by keeping it damp humidity can be controlled.

It is very important that the substrate be kept clean. All uneaten food and remains of food should be removed as soon as possible. It is virtually impossible to create an environment completely free of mites but these will not be a problem if cleanliness is maintained. They will multiply at an astonishing rate on dead material, particularly in the hot, humid environment of the tarantulas housing. By maintaining cleanliness the spider need only be disturbed a couple of times a year, when the whole container and contents should be cleaned thoroughly. It is a good idea when cleaning or moving the spider to a new container to retain a small piece of the web and place it on the fresh substrate, or bark if an arboreal species. The tarantula is less likely to become distressed and will settle more quickly with its own scent in the container.

**Furnishing the container** The captive tarantula has no interest in decorative additions to its home. As long as the basic requirements are met, the decor is for the benefit of the owner. If the spider is to be on show the set-up can be made to look quite attractive and realistic.

The basic requirement is a retreat. Tarantulas generally do not like to feel vulnerable. In their natural habitat, in a burrow, in their web or under a rock, they will be in a very confined space, spending the majority

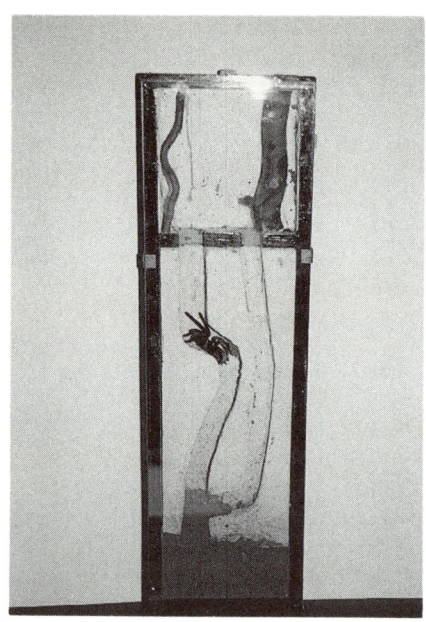

Fig. 17 Housing constructed from a building block. *Haplopelma minax* in burrow.

Fig. 18 *Lasiodora parahybana* in an aquarium.

Fig. 19 Juvenile *Euathlus smithi* in an aquarium.

Fig. 20 A custom made container for *Pterinopelma saltator* (The White Collar).

of their time in safety. In a bare container they will retreat to a corner. Something should be put in the container under which the spider can hide. Cork bark is a natural material which can be obtained in a variety of shapes, some suitable for the tarantula's retreat. It will be able to fix its web easily to the bark and will venture out at night once it has settled in. Other types of bark can also be used, having been sterilized first. A flower pot laid on its side makes a good retreat. Square ones are more stable and plastic is more easily cleaned, although light enough for the spider to move if it so wishes. Another natural retreat is a coconut shell, a 'spider sized' hole having been cut and the coconut removed. A flat piece of stone placed on the Vermiculite and a space cleared under it will suit the species which do not make a burrow, but a scrape under a rock. Care must be taken that the stone will not fall on the spider.

Individual tarantulas behave in different ways with regard to their retreat. The carefully placed flower pot may not be to the spider's liking so it may move it. It may prefer to sit on top of it or behind it so there should be no sharp edges. Tarantulas which make a lot of web such as Asian species or Baboon spiders often use the retreat as an anchor point for the web. At moulting time when the spider will be at its most vulnerable it may go into its retreat and web across the entrance. If this happens the spider should be left alone and it will emerge a few days later having shed its skin in safety. In the wild it would behave in a similar way.

Plants can be a colourful addition to the housing but, as said earlier, cactus plants can be lethal therefore should never be used. Plants which are used must be tolerant of the high temperature and humidity required by the spider, so, to avoid any problems, artificial plants may be used instead. There should be no sharp points onto which the spider may fall. Sphagnum moss helps to keep the humidity level up through transpiration of the leaves. A piece can be placed in a corner and renewed when necessary. Arboreal tarantulas will need vertical bark or twigs on which to construct their web. These should be sterilized before use and renewed when the container is cleaned. An arboreal set-up lends itself to the use of plants.

Young tarantulas of arboreal species will not need a full sized container. They often live under bark, so to be able to observe them use two plastic containers of different sizes. Place the smaller one inside the larger one with substrate between them. The spider will happily live between the two containers and use the centre for hunting. This method has been successful in rearing *Poecilotheria regalis*, the Indian Ornamental.

Chapter Five

# Feeding

Tarantulas, along with other arachnids, are predators. Unlike scavengers which will take dead food, these predators must kill their own food. Amongst arachnids some species actively hunt, others lie in wait for prey to pass and some build elaborate snares. The natural diet of a wild tarantula will consist largely of invertebrates, but they will also take small mammals and reptiles. The arboreal species may occasionally take small birds hence the common name 'bird eater'.

The majority of tarantula species are sedantary creatures which do not move more than a few feet from their retreat (burrow or web) therefore rely on the prey passing close by. In captivity allowances must be made for this by providing living quarters of a size whereby the tarantula can feed easily without being unduely annoyed by unwanted food. In the wild a tarantula may be lucky enough to catch passing prey frequently during the course of a few days, then nothing for weeks. Due to its ability to store predigested food within its body this does not present a problem.

The captive tarantula will be provided with a regular source of food and it should be established what the eating habits of an individual are – will it eat every day or once a week? Certain species will eat frequently, other species rarely, but even within a species there are variations between individuals. It may take as long as a year to establish a feeding pattern as a tarantula will stop eating before a moult. This period varies depending on the age of the spider and the species. Periods of six or seven months are not unknown for an adult. Accurate records should be kept, recording date of feeding, food used and whether the food was eaten or not. If the dates of moulting are included a pattern should emerge. It can be worrying if a tarantula refuses to eat. If the temperature and humidity are correct it could be that it is approaching a moult or is a particularly inactive species. Well-kept records will dispel this worry once a pattern has been established. (Fig. 21)

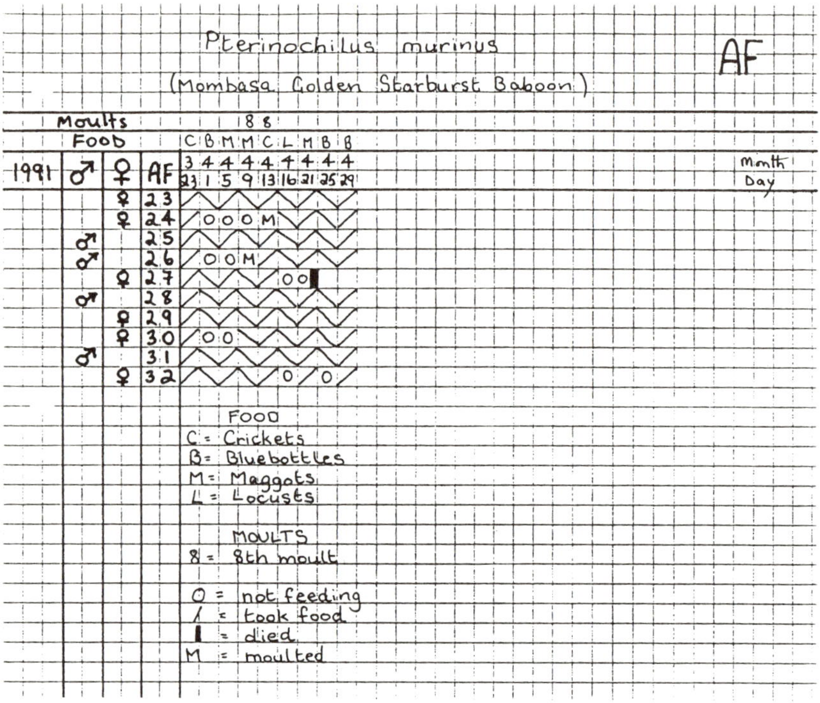

Fig. 21 Feeding record.

Some species, generally those that are more aggressive such as Baboon spiders, will pounce on the food as soon as it is introduced. Other species prefer to eat at night. Uneaten food should be removed after 24 hours whether it remains alive or the spider has killed it. If a wild tarantula does not want to eat it will let the prey pass by whereas in the confined environs of the captive tarantula live food may not only be worrying but an actual danger to the spider.

Approaching a moult (see 'Moulting') a tarantula will stop feeding. It is when this occurs that care must be taken not to leave uneaten food in the container. Particularly dangerous are crickets as they have been known to try to eat the tarantula when it is at its most vulnerable. As it emerges from the old skin it is still soft and defenseless. Small spiderlings easily fall prey to crickets at this time.

Choice of food is important. It must contain no venom which could harm the tarantula and not be carniverous itself. (Crickets are omniverous, but do not harm a healthy spider that can defend itself.) Wasps and bees should never be used as food and certain caterpillars can cause irritation with their hairs. Beetles should be selected carefully as some species have biting jaws which could harm the spider. Leaf eating beetles and mealworm beetles, although not a popular food, can be used if necessary.

The size of the prey depends on the size of the tarantula. An adult of the popular species can easily deal with an adult locust. With immature spiders and the smaller species the size of the prey should not exceed the body size of the spider. It is surprising how large an item of prey a tarantula will tackle but care should be taken that the prey will not harm the spider. Locusts have spines on their legs when adult and these can tear a tarantula's abdomen. Locusts and crickets are the most popular food as they are usually easily obtainable from pet shops. These can be used in all stages of development to suit the size of the tarantula. A variety of food will ensure that the spider has a balanced diet.

It must be considered at all times that the captive tarantula is in an artificial environment no matter how carefully its housing has been thought out. For this reason unwanted food should be removed. Dead remains should not be left as they will encourage mites and mould which could eventually harm the spider.

The mechanics of feeding in tarantulas is similar to that of all arachnids. The food is digested outside the body. This can be observed whilst watching a tarantula eat. Its primary task is to immobilise the chosen prey. With its downward pointing fangs it must strike the prey from above and against a solid surface. This is done by raising itself on its forelegs, opening its fangs and striking. This process takes a fraction of a second. Once the fangs have entered the prey venom is injected. The prey is now immobilised . This may take a few minutes for the venom to take effect. The tarantula can begin to feed. It can only absorb food in a liquid form so it excretes digestive juices from its stomach out of its mouth onto the prey. It then proceeds to mash the food, with the help of its fangs, pedipalps and chelicerae, into a pellet. It sucks the resulting liquid into its stomach. Any remaining particles are sieved by the orange hairs which act as a filter and by the rostrum plate (small overlapping plates which only allow pre-digested liquid through). The tarantula will continue to manipulate the remaining pellet until either it has eaten enough or only indigestible material remains. The resulting pellet is then discarded and can be removed to ensure cleanliness in the container.

**Locusts** Locusts are used as a food source for a variety of captive species, from lizards to mammals as well as for arachnids. They are therefore not difficult to obtain from pet shops and by mail order. They can be used in all stages of development, from hatchlings (hoppers) through to adults complete with wings depending on the size of the tarantula. Adult locusts have sharp spines on their rear legs which can cause damage to a tarantula if it kicks. If the locust is 'served' at room temperature it will be quite subdued. The optimum temperature for locusts is 32°C (90°F) so even in the spider's container they will be relatively inactive. Large hoppers, in their 5th instar do not present such a problem and are generally more palatable to the spider as they have not got such a hard exoskeleton and have no wings which the spider often discards anyway.

If locusts cannot be bought singly it is quite easy to keep them for at least a couple of weeks. They can be kept at around 28°C (82°F) and fed on fresh grass. If no grass is available they will eat bran but in this case they must be given water. They will not survive if left with the spider as feeding them will cause problems and their presence will annoy or even damage the spider. A plastic sweet jar can be used, with twigs as branches. If the jar is inverted the lid can be removed without the locusts escaping and droppings can be discarded and food put in. Holes should be made in the plastic to prevent a build up of humidity. (Fig. 22)

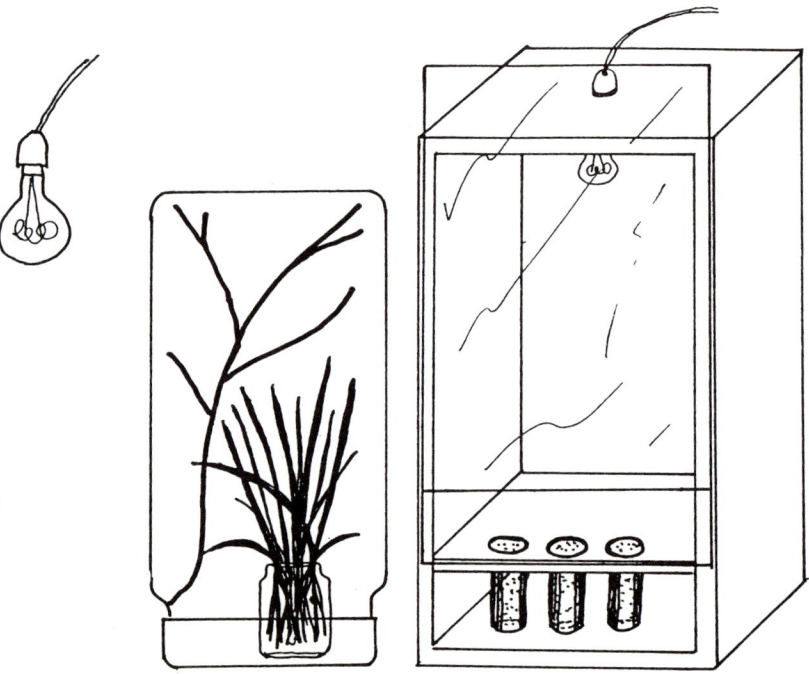

Fig. 22 A holding jar for locusts.     Fig. 23 Locust breeding set-up.

For the enthusiast who has built up a collection of tarantulas, the answer is to breed the locusts. This will provide a continuous supply of food of a variety of sizes and reduce feeding bills considerably. The ideal size of container in which to keep about 40 breeding pairs of locusts is 40cm. × 40cm. × 50cm. high. This can be made of plywood, hardboard, aluminium sheeting or perforated zinc with a glass front. If a false floor is fitted, made of perforated zinc, the droppings can fall through the holes to facilitate cleaning. The holes in the zinc should be small enough to prevent the newly hatched hoppers falling through. Heat can be provided using light bulbs and can be controlled by varying the wattage and monitoring the temperature given by various sized bulbs. (Fig. 23) Locusts need both heat and light for at least 8 hours a day and preferably 12 hours. The daytime temperature for successful breeding is 32°C (90°F) with a drop to 28°C (82°F) during the night. They will survive at lower temperatures but the reproductive cycle and growth rate will be slower. This may be advantageous if only limited numbers are required.

Provision must be made for egg laying. In the wild locusts lay their eggs in the ground, to a depth of up to 10cm. Clean, tightly packed, damp sand in a plastic or glass container about 5cm. in diameter is suitable. If the cage is fitted with a false floor and the tops of the tubes are level with this floor, they can be removed and replaced easily. They should be removed every day or two because too many egg pods close together will inhibit hatching. When removed the containers should be covered to prevent the sand drying out, and kept in a warm place (28°C to 32°C) until hatching is due two to three weeks later. They can then be placed in, preferably, another breeding cage as cannibalism often occurs. Locusts do not have to have specialised housing but can be bred successfully in a covered aquarium with sand containers placed on the bottom. There should always be twigs in the container as the locusts will hang from these whilst moulting. Hygiene is important. The cage should be cleaned regularly and twigs discarded and replaced with new ones. Cleanliness should prevent disease which could be passed on to the spiders. Locusts are liable to pick up infections of nematode worms from grass. If this happens, change the source of the grass.

The two species of locust most commonly kept as food are the African migratory locust (*Locusta migratoria*) and the desert locust (*Schistocerca gregaria*). These species will thrive on a diet of fresh grass, which should be replaced every day, and bran. No water is required if fresh grass is available as it is in Britain throughout the year. If the grass is compressed into the neck of a glass jam jar with most of it protruding it will stay fresh for longer at the high temperatures and the uneaten remains can be removed easily. Alternatively wheat can be planted in shallow trays and when it is about 10cm. tall the tray is placed in the locust cage. The disadvantage of this is that the locusts will lay their eggs in the tray. Locusts can be fed on an artificial diet made up, by volume, of one part bran, one part dried milk, one part dried grass and one tenth part dried yeast. If this is used water must be available. To prevent the hoppers drowning, cotton wool soaked in water should be used.

For further information on the breeding of locusts a booklet is available from the Anti-Locust Research Centre entitled 'Rearing and Breeding Locusts in the Laboratory'.

**Crickets** Crickets are also a widely used food species which are easily obtainable from pet shops and by mail order. The species most commonly available is the house cricket (*Acheta domestica*) but the black African field cricket (*Gryllus sp.*), which is becoming more popular, grows much bigger so is more suitable for adult tarantulas. House crickets, if they escape, can be quite a pest, fouling food and chirruping incessantly in inaccessible places in the home. The larger field crickets are slower moving and easier to handle. They are not such a problem if they do escape. Both species of cricket can be used at all stages of their development. Newly hatched micro crickets can be fed to spiderlings through to adult crickets for large spiders. However, where one locust might satisfy a tarantula, a number of crickets will be required, particularly if the smaller house crickets are being used.

A few species of cricket are purely vegetarian, but many are omniverous. For this reason care should be taken when using them as food because they can damage or even kill a tarantula. This is particularly so with the house crickets which have been known to start to eat a tarantula whilst it is moulting. Accurate records of the tarantula's moulting and eating habits will help to avoid this happening. If it is coming up to moult or reluctant to eat, don't use crickets. Crickets can also annoy a spider which is not eating if left with it too long. They should be removed after 24 hours.

Because crickets are fast moving they may be difficult to catch, both by the spider and the owner. To slow them down before feeding, they can be cooled slightly in a refrigerator. Micro crickets, due to their size, are not easy to pick up. This can be overcome by using a pooter. The cricket is sucked up, stops at the filter at the centre of the tube, the end of the tube is placed in the spider's container and the cricket is gently blown out. In this way the number of crickets given to a spider can be controlled and accounted for if they have to be removed. The pooter can be used without causing too much disturbance to the spider. (Fig. 24)

Crickets can be kept for some time if not all consumed at once. They

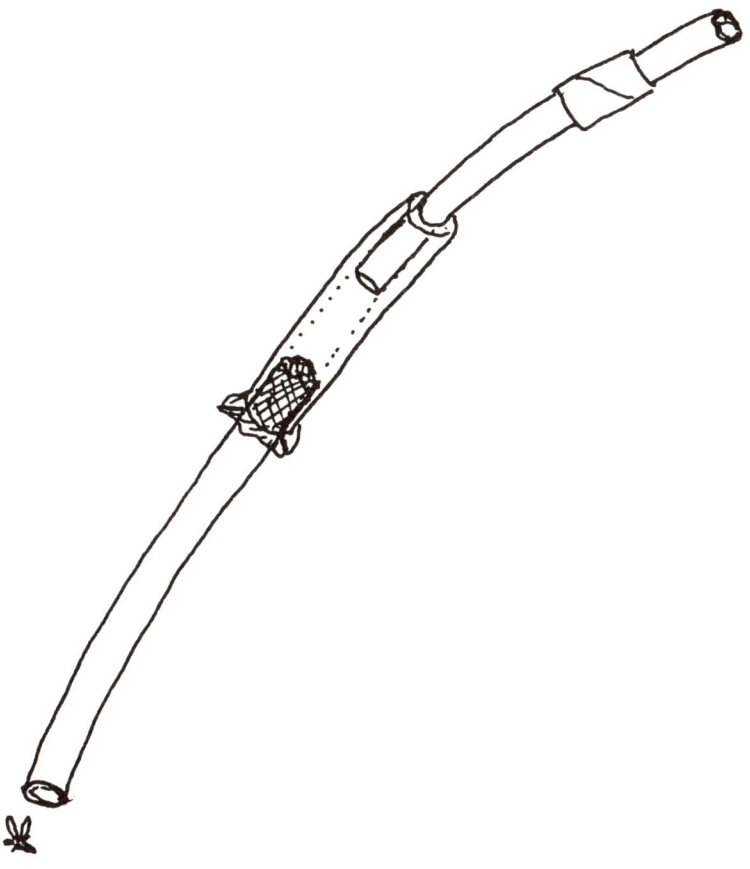

Fig. 24 A pooter.

should be placed in an aquarium or other suitable container with a close-fitting lid made of fine mesh. Metal mesh is preferable as they can eat through material. They should have a large surface area on which to move and this can be created by putting egg trays or boxes, crumpled newspaper or a roll of corrugated card into the container. Crickets will eat almost anything, including the card, and can be fed on dry household scraps such as cornflakes, dry bread, dog or cat biscuits, small amounts of vegetable peelings and lettuce. It is unwise to give them too much green food as the humidity level will rise and they will die. High humidity will also encourage mites which could be transferred to the tarantula. Water can be provided by soaking a piece of cotton wool in water and placing it in a small dish.

The container should be cleaned out frequently, both to keep the crickets healthy and prevent a build-up of mites which could harm the tarantula when it is fed. The water container should be replaced every day. Breeding crickets can be done successfully on a small scale. Their life cycle takes longer than that of the locust but it can be speeded up by increasing the temperature. At 20°C to 25°C the life cycle of the house cricket is 3 to 4 months. With the temperature increased to 30°C to 35°C this can be reduced to 5 to 6 weeks. Heating can be provided by a light bulb but if this is connected to a thermostat other lighting, for example fluorescent, must be on for a recommended period of 16 hours per day for the best results. This can be connected to a time switch. If crickets are not provided with a suitable medium in which to lay their eggs, they will do this in the food dish, water dish or substrate. Crickets normally lay their eggs in the soil, singly or in batches. A plastic flower pot filled with moist sterilized peat or sand can be replaced about twice a week to avoid too many eggs being layed in a small area. This should result in a steady supply of hatchlings. Adult crickets may eat newly layed eggs so the top of the container should be covered with metal gauze through which the females can put their ovipositor and lay their eggs. Nymphs of several batches can be kept together but it is unwise to add first instar nymphs to a container with fourth instar or larger crickets as the larger ones can be cannibalistic.

**Flies** Flies are often overlooked as food for tarantulas, probably because they are small in relation to adult spiders. However, they are ideal for spiderlings and sub-adults as well as providing a change of diet for adults. These may take as many as a dozen at a time. Arboreal species seem to prefer flying prey. Flies are not as difficult to control as may be first imagined.

The easiest form in which to use flies is when they are larvea or maggots. Four types are available and suitable for tarantulas. These are Bluebottles or Blow flies (*Calliphora erythrocephala*), Greenbottles (*Lucila caesar*), Houseflies (*Musca domestica*) and Fruitflies (*Drosophila melanogaster*). The first three types can be bought from fishing tackle shops and fruitflies can be obtained from specialists in live food or cultured or collected at home. Maggots intended for fishing are sold as 'Whites' (Bluebottles), 'Pinkies' (Greenbottles) and 'Squats' (Houseflies) and can be bought in small quantities. Dyed maggots should not be used, neither should the flies which emerge from them, as the dye may harm the spider. Although maggots should not be fed to reptiles and amphibians which swallow them whole, they cannot damage a healthy spider. The spider digests them outside its body so there is no possibility of damage to the internal organs. Maggots will not attempt to eat a living creature, so will not try to feed on the tarantula. They are in fact rather particular about what they eat.

Once the maggots have been obtained they should be kept cool to retard pupation. This can be done by keeping them in a refrigerator where the temperature does not fall below freezing point. When the maggots are fed to the tarantula many appear to enjoy scratching about in the substrate to catch them. Uneaten maggots do not pose a problem because they will burrow into the substrate and pupate. When they emerge from the pupae, the spider has another meal, this time flies. This system can be put to good use when the owner is going on holiday. The spider can be fed three meals at once, flies, pupae and maggots. It will eat the flies first, about a week later the pupae will hatch, followed by the maggots which had in the meantime pupated. The number and species of flies or maggots can be varied according to the size and appetite of the spider. Apart from irritating the spider, flies will not cause it any harm, but nevertheless should be removed if uneaten or dead.

If pupae are fed to the tarantula this is the easiest way to handle flies. The maggots will pupate if left in a warm place. The pupae gradually darken in colour, so the darkest are nearest to hatching. For ease of handling flies, the following method is suitable. Take a container such as a 4 litre ice cream box. Cut large holes in the lid, removing as much material as possible, but keeping the lid fairly rigid. Place a piece of nylon netting over the top of the box and replace the lid. Cut a small hole in the netting and plug with a piece of foam or sponge. This hole allows a pooter to be inserted to catch the flies. Mix some maggots with sawdust or Vermiculite so that the bottom of the box is covered to a depth of about 2cm. If left in a warm place the flies will emerge in one or two weeks. Feed the flies by putting a small amount of sugar water on the netting. By using a pooter the lid need not be removed, so avoiding escapes.

Fruitflies (*Drosophila sp.*) can be used to feed small spiderlings. During the British summer these will be attracted by fruit such as banana or orange left in a sunny place. They can then be collected using a pooter. If greater quantities are required, a culture can be set up for a constant supply. It can be started by attracting wild flies as above or a culture can be obtained from a live food supplier. The following method, once the colony has been set up, will keep it going indefinately if a fresh mixture is prepared once a week.

Ingredients:- 2 teaspoons dried yeast
100 ml water
½ teaspoon malt extract
Bran
1 orange
3 drops Nipagin (to prevent mould)

Method:- Put 1 tsp yeast in water and boil for 15 minutes.

Finely chop orange.
Take yeast water off the boil and add orange.
Stir in Nipagin.
Mix with bran to make a thick paste.
Place in container and sprinkle on rest of yeast.
Stand in a warm place.

If some flies are put into the container, the culture will continue. It can be kept in the same way as for the larger flies. Maggots, pupae and adult flies can be used for feeding.

Due to their size fruitflies can be difficult to contain and can be annoying to other members of the household, particularly as they are attracted to alcohol. A wingless variety has been bred, but it is difficult to culture because a winged fly mating with a wingless one will revert to the natural variety. A pooter is invaluable when using fruitflies as food.

**Cockroaches** The tarantula keeper who also has other invertebrates may find he has a surplus of cockroaches. These are harmless to tarantulas and some of the larger species are easy to rear and are quite prolific. The American cockroach (*Periplaneta americana*) can be quite a pest in homes if it escapes so should be avoided. More exotic species from tropical regions would not survive in temperate latitudes and, particularly the flightless varities, are easy to keep and rear. The Cuban Burrowing cockroach (*Byrsotria fumigatata*) has no inclination to climb the sides of its container and, although the males have wings, they cannot fly. They breed quickly and the nymphs can be used for smaller tarantulas. To breed them put a deep layer of peat and some bark in a covered aquarium and feed them on household scraps such as bread, currants, cereals and some vegetables and fruit and leave them to get on with it! They need only be fed once a week. Hissing cockroaches (*Gromophadorhina sp.*) have a longer life cycle but the adults are much larger. They can be kept in a similar way to the burrowing cockroaches, but do not need such a deep layer of peat. Cockroaches should be cleaned out regularly as they can carry mites which may be transferred to the tarantula's quarters.

**Stick Insects** (*Phasmids*) Some tarantulas will feed on the Indian Stick Insect (*Carausius morosus*) which are easy to rear. They do not require heat and can be fed on privet or bramble, both of which can be found in winter. The ova hatch soon after laying and one stick insect can produce a very large number of these. Many species are parthenogenetic so no male is required to fertilize the eggs. Nymphs through to adults can be fed to tarantulas according to size. Surplus eggs can be fed to goldfish, etc. Some species of phasmid have sharp spines and should not be used as food.

Although the Indian stick insect is a convenient source of food, some tarantulas are reluctant to eat them.

**Worms** A good standby in winter when the ground is wet is earthworms. They provide a good source of nourishment and many of the larger species enjoy them. They should only be given to spiders which will eat as soon as it is introduced otherwise the worm will burrow into the substrate where it will die. Nothing smells quite as bad as a dead worm. Any uneaten pieces should also be removed as soon as possible. A large tarantula such as the Goliath (*Theraphosa leblondi*) or the South American Bird Eaters (*Pamphoboetus sp.*) will break a worm up. Small pieces are dropped by the spider and these decay quickly in the high temperatures.

**Moths and Caterpillars** Many tarantulas, particularly arboreal species such as *Peocilotheria regalis*, the Indian Ornamental, will feed on moths. Even in Britain some species are quite large. Smaller species can be fed to juveniles and sub-adults.

The most efficient method of catching moths is by using a mercury vapour light and moth trap at night. The light attracts the moths which fall into, for example, a plastic bucket on which the lamp is sitting. If crumpled newspaper is put into the bucket the moths will stay amongst this and can be removed during daylight hours. If a moth trap is not available, a simple method of attracting the moths is to leave a window open and a light on. Caterpillars will provide alternative food, but great care must be taken when using them. Certain species have irritating hairs as a defense and may damage the tarantula. Only hairless species should be used but only if they have not come into contact with horticultural or agricultural herbicides or pesticides. If in doubt, don't use them.

Waxworms are the larvae of a moth and these are sometimes available from suppliers of live food. The larva or moth can be used as food and there is no danger of these having been contaminated.

**Mealworms** Mealworms are a good standby food in an emergency. They are not very nutritious so should not be used on a regular basis. They can be cultured easily by putting them into a container half filled with bran. Put a slice of bread on top and give them a lettuce leaf about once a week. The larvae will pupate and hatch as beetles which can also be used as food as they are harmless. The beetles will lay eggs which will hatch as very small mealworms which can be used for small spiderlings. The culture will continue as long as fresh bran is added at intervals. A giant species of mealworm, sometimes called a Morio worm is also available from live food suppliers. Mealworms will burrow into the substrate if the spider does not eat them immediately. They should be removed if they do this. Even large tarantulas will take mealworms, eating six or so at a time.

**Vertebrates** The tarantula's diet does not have to be restricted to invertebrates. In the wild they will take small mammals and reptiles. Pinkie mice and gerbils (i.e. newly born) provide a substantial meal for an adult tarantula. These can often be obtained from pet shops that supply live food for snakes and other reptiles. If a large number are required they can be bred quite easily. Lizards also give a nourishing meal. This may appear to be a very expensive way of feeding a tarantula, but those owners who live in warm climates should have little trouble catching or breeding them.

**Dead food** Some tarantulas have been persuaded to take dead food in the form of meat if this is dangled on a thread in front of them. This is an unnatural practice and should be avoided, but it has been used to persuade an ailing tarantula to feed.

To sum up on feeding, the food must be alive, pose no threat to the tarantula and be of a suitable size. In the wild the spider will be able to ignore or escape from potentially dangerous prey, although a Goliath Bird eater was observed killing and eating a fer-de-lance, one of the most deadly snakes in the world. Within the confines of its housing the captive tarantula will not be able to escape from danger. Also in these artificial surroundings, and particularly with captive bred specimens, some of the natural instincts may have been lost. The golden rule of feeding is – if in doubt about a species of food, don't use it.

Chapter Six

# Moulting

The tarantula, in common with many invertebrates, has a rigid exoskeleton which restricts its growth. Only the soft abdomen can expand, so to allow for growth it must shed the entire skin. Everything on the outside is shed – all the hairs, the spinnerets, claws, lenses of the eyes and some parts of the inside too, namely the book-lungs and the spermathecae. Even when the tarantula is an adult female it will continue to moult at regular intervals, usually annually. As well as growing slightly, it has the opportunity to repair damage to its body.

Moulting is a process of two different events. First, a gap is created between the old layer of skin and the new by enzymes dissolving away certain tissue. This leaves the exuvial space and cells then begin to secrete the new cuticle. This process is known as apolysis and at this time the adult tarantula will become very quiet and web itself into its retreat. The actual process of shedding the skin is known as ecdysis. At this time the spider is particularly vulnerable and should not be disturbed. It can be damaged very easily.

There are signs to look for when a tarantula is approaching a moult. The first is that it will stop feeding. The slower growing species may refuse food for a month or more, one example being an *Apholopelma seemanni*, Costa Rican Zebra, which did not eat for seven months. This can be worrying for an inexperienced owner, but the tarantula does have the ability to store food within its body and all its energies must be put into the complicated process of apolysis. An obvious sign that the tarantula is approaching a moult is the darkening of the skin. This is particularly noticeable where the spider has a bald patch on its abdomen caused by throwing hairs. This patch will gradually change from being quite pink to black. This darkening is not the old skin, but the hairs on the new skin showing through. The abdomen will also have the appearance of approaching bursting point, particularly in a well fed specimen. This fact causes the legs to appear thinner. Another sign is the movement of the abdomen away from the prosoma so that the pedical is visible. (Fig. 25)

Unless the tarantula has hidden itself in its retreat the next stage of the process can be watched. It will lay some extra web on the substrate and then lie on this web on its back. The spider is not dead. Dead tarantulas do not lie on their backs. On no account should it be interfered with. Once the spider is on its back the process of ecdysis begins. A split occurs at the front of the carapace and continues down the sides. The spider can be seen to move its chelicerae to aid this. The carapace, complete with the lenses of the eyes lifts like a lid. The tears on the prosoma now continue to the opisthosoma and the skin of the abdomen splits at both sides. Once this has happened the spider can begin to free its extremities. It will slowly move its legs and palps gradually pushing the old skin upwards. The new hairs prevent it from slipping back. When all the extremities are free the tarantula will cast the old skin to the side and eventually return to an upright position. (Fig. 26)

The new skin, particularly in an adult, may take a few days to harden. On moulting the fangs are soft and appear white, as do the leg joints. The tarantula is virtually defenseless, hence no food should be introduced and it should not be touched or disturbed. Because the new skin takes time to harden it is important that the spider has enough space for this to happen in a natural position. If cramped in a container that is too small, its legs may be deformed, causing difficulties at the next moult.

As long as it does not disturb the spider too much the old skin can be removed. It will be wet if this is done immediately and can be set in a natural position. All that will appear missing will be the abdomen which is a thin skin which has collapsed. This is an excellent opportunity to examine the anatomy as everything on the outside of the spider will be on the old skin, hairs, fangs, chelicerae, spinnerets, legs, carapace, etc. Identification of species can be carried out without having to have a dead specimen. The skin is, in fact, an exact replica of the living spider. Sex can also be determined at an early age from a shed skin and this is important if a breeding programme is being carried out. (See 'Sex Determination')

Sometimes, particularly in spiderlings, a patch of old skin remains stuck to the abdomen. this can be removed gently with a soft artist's brush dipped in water. Attempting to pull the old skin off will damage this new one, perhaps splitting it, with fatal results. A cause of this is if humidity is too low. At moulting it is best to have the humidity erring on the high side. The whole process of moulting takes longer as the tarantula gets older. Spiderlings may only refuse food for a day, and may not show any visible signs, whereas an adult female will take much longer to complete the

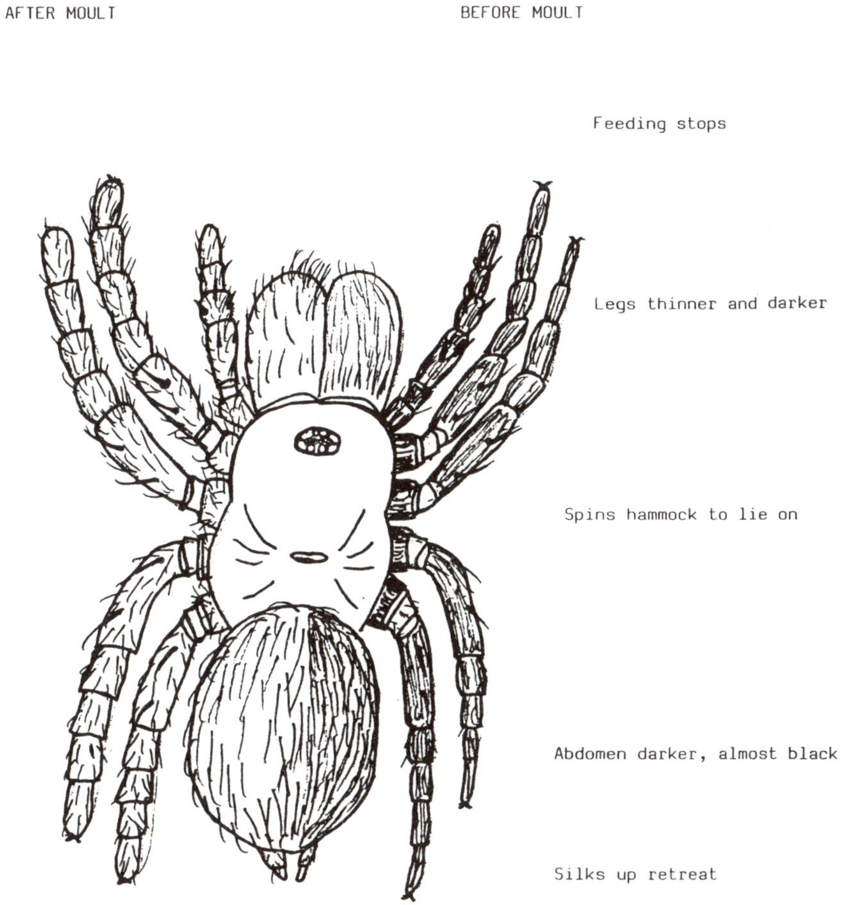

AFTER MOULT                    BEFORE MOULT

Feeding stops

Legs thinner and darker

Spins hammock to lie on

Abdomen darker, almost black

Silks up retreat

Fig. 25 Predicting an approaching moult.

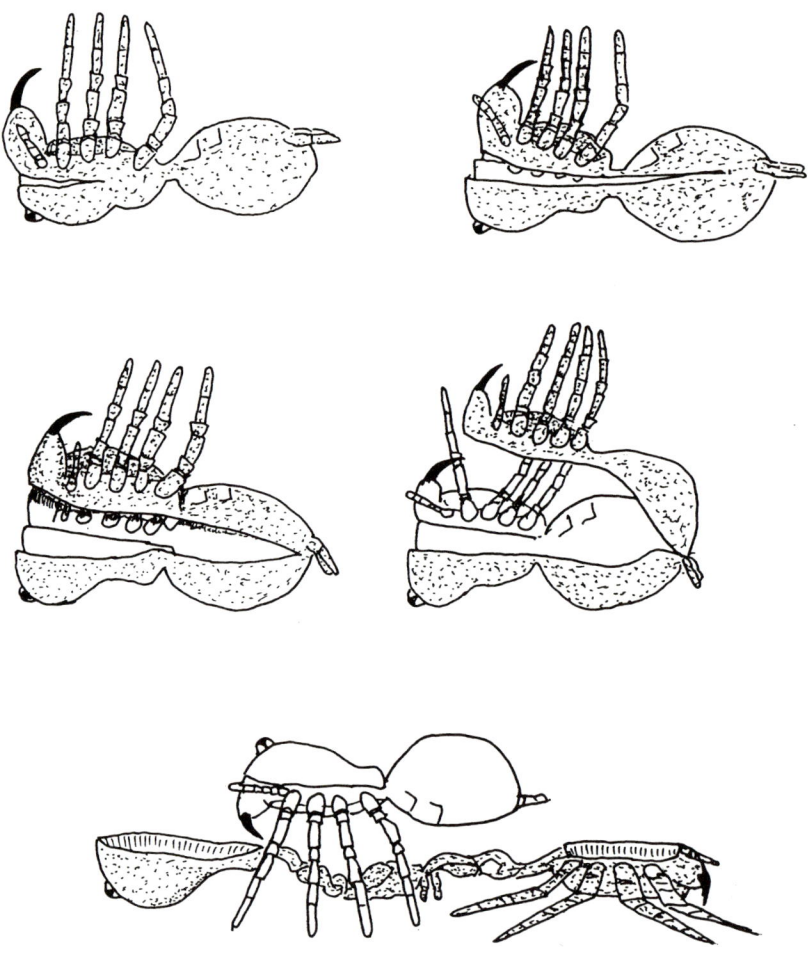

Fig. 26 The process of ecdysis or moulting.

process. Adult males will not moult. Once they have reached maturity their purpose in life is to find a female to mate with and they die within a few months. This prevents inbreeding as a male can never mate with his offspring or with his siblings as he will have matured and died by the time his 'sisters' have reached reproductive age. It has been reported on a couple of occasions that a mature male has attempted to moult, but the mature palps which are enlarged have become stuck in the old skin and he has died.

Apart from allowing it to grow, moulting enables the spider to regenerate lost limbs. It is well worth buying a tarantula with a missing limb and a bald abdomen at a reduced price. After moulting it will be complete. Sometimes the regenerated leg of an adult may be thinner than the others but a second moult will put this right. Tarantulas, like the true spiders, are able to detach a leg if it is caught by a predator or trapped. This is called autotomy. It will happen if a leg cannot be extracted from the old skin. This occurs if the atmosphere in the container is too dry and the lubricating fluid between the old and new skins dries up too quickly.

Observations have shown that tarantulas are, to some extent, able to control when they moult. This is important with mated females which can delay moulting until the egg-sac has hatched. Spiders which have lost up to four legs have been observed to moult shortly after the previous one, and reappear complete. Females which are getting old will moult soon after mating to prevent the development of fertile eggs. Young females which are not quite mature will also moult early. When a female moults it becomes virgin again, so the mating was useless. The whole process of moulting is very complicated and takes a lot of energy on the part of the tarantula. All the owner can do is wait, having ensured that the humidity level is erring on the high side to prevent the old skin from sticking. The freshly moulted tarantula will be immaculate if all goes well. If the date is recorded, the next moult can be predicted.

Chapter Seven

# Sex Determination

To be able to breed tarantulas it is obvious that the sex must be known. This is not as simple as it might appear, because, particularly with a wild caught specimen, there may be no outward indication as to whether it is male or female. This is because sexually immature theraphosids cannot be sexed at a glance as both sexes appear identical. Neither is there a positive indication that it is mature in the majority of cases. Many a prospective breeder has been disappointed to find that the 'mature female' he has just purchased moults into a handsome mature male. The only positive indication of sex and maturity at a glance is if the female is carrying an egg-sac or a male has fully developed palps and, in most species, mating hooks on the first pair of legs. If this evidence is not present the tarantula could be an immature of either sex. The same problem arises with rearing spiderlings with the prospect of breeding. If the sex is known at an early stage, the time at which males mature can be regulated by feeding to coincide with the best time to mate an available female. Sexing also avoids the potentially disastrous results of putting a mature male with what is thought to be a mature female, which is in fact another male although immature.

The sexual organs of a spider open at the epigastric furrow, present in both males and females. From the outside an expert may be able to ascertain the sex by feeling this area. The spider must be picked up so that a finger can feel along the mid-line between the two pairs of book-lungs. If there is a distinct bump or nose above the epigastric furrow, the specimen is most likely to be a mature female; if the area is flat it is probably a male. (Fig. 27) The size of the protrusion varies greatly between species, so this method is not reliable. The only positive means of determining the sex of a living tarantula is by examining the shed skin. To do this it is necessary to have an understanding of the internal anatomy.

In the male spider the testes consist of two long, greatly coiled tubes which

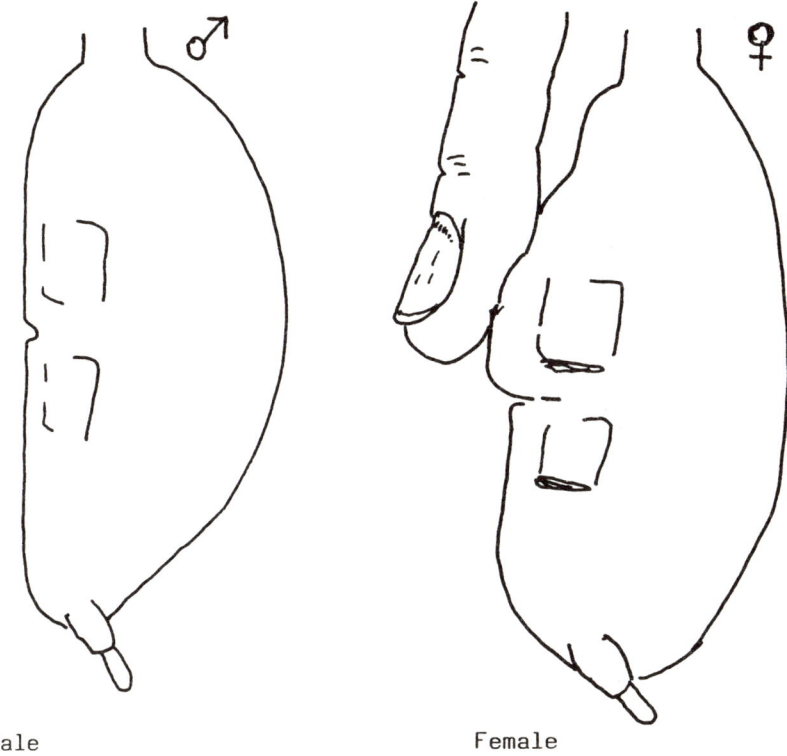

Male                        Female

Fig. 27 External sex determination.

function as both vas deferentis (sperm ducts) and the site of spermeogenesis (where gamete cells differentiate into sperm cells). These tubes open to the exterior via a single opening known as the gonopore. Externally the gonopore can be found beneath a long fold of skin between the anterior book-lungs, the epigastric furrow. Internally the gonopore opening is often encircled by two thick 'lips'. (Fig. 28)

The female's reproductive organs consist of a pair of ovaries which, at times, can be enlarged like a bunch of grapes. Theraphosids' ovaries are similar to those of birds in that the oocytes (further eggs) bulge out into the body cavity. The ovaries are connected to the exterior by tubes, the oviducts. These open into the uterus externus, an enlarged short section of tube more like a pocket than the junction of the two oviducts. From this pocket, branching off into two blind sacs, are the spermathecae. It is into these that the male injects his sperm for storage. The female's

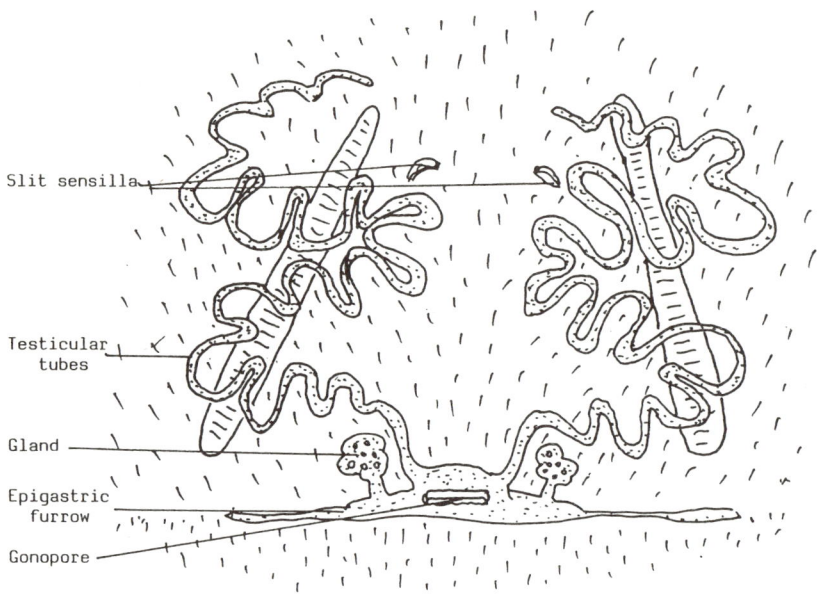

Fig. 28 Male Theraphosid sex organs.

reproductive system opens to the outside via a gonopore which, as in the male, is located within the fold of the epigastric furrow. The spermathecae are lined with cuticle, as is the uterus externus pocket. In the theraphosids which moult even in their adult stage, this lining is shed with the old skin. This means that any previously mated female becomes virginal again because any stored sperm will be lost with the moult skin. The presence of spermathecae establishes the fact that a spider is a female. (Fig. 29)

In the larger species of theraphosid it is often possible to make out the form of the spermathecae without magnification, and smaller ones by using a X2 magnifying glass. A X10 lens can be used from the eighth moult onwards, but for smaller species and younger spiderlings a microscope is necessary. Stereoscopic binocular instruments are the best type to use with magnifications from X5 to X50. Other equipment needed will be two pairs of very fine forceps, one curved and one straight, a pair of fine dissecting scissors, a sharp scalpel and two mounted needles.

To sex a moult the skin must be soft and flexible. The best method is to place the skin, for a few hours, in a jar containing four drops of washing up liquid to 500ml. water. The leg region will contain air so will float on the surface. The solid carapace will sink and pull at the twisted rope of skin

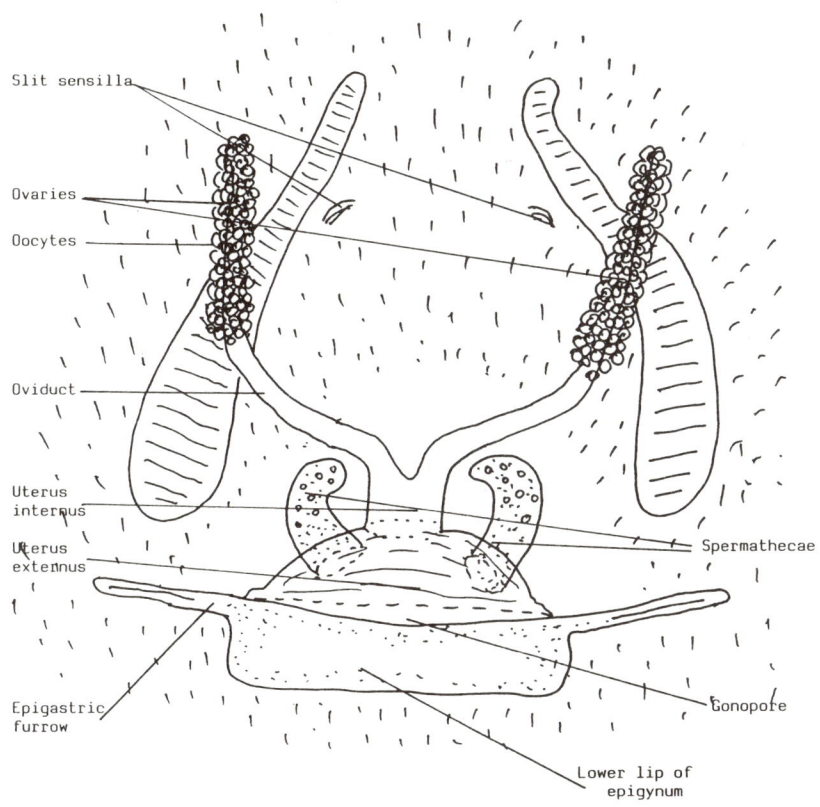

Fig. 29 Female Theraphosid sex organs.

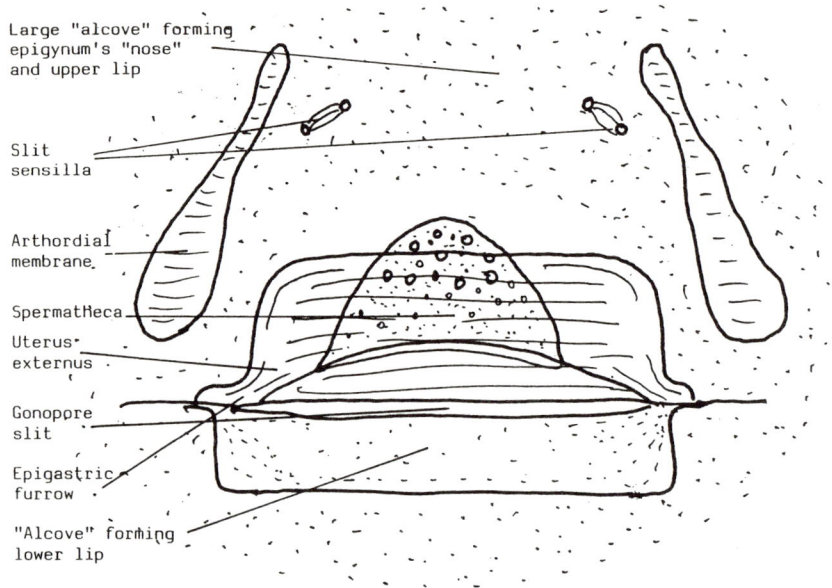

Fig. 30 Female epigynum showing spermathecae (viewed from the inside).

that once covered the abdomen. Now place the softened skin in a petri dish, just as a living tarantula would sit and taking great care not to tear it, untwist the abdomen skin. Gently unroll the edges to reveal the book-lungs which will appear white. Between the book-lungs is the area to be examined. It is between the anterior pair that an alcove will be observed. Next locate the slit sensilla which are specific stress receptors which look like minute button-holes. Directly below these organs will be found the spermathecae, if the specimen is a female. (Fig. 30) Spermathecae come in two main types, fused or paired. (Fig. 31)

The structure needs to be understood to differentiate between male and female. The path of the male embolus as it enters the female copulatory organ must be followed. (Fig. 32) First the embolus will seek out the epigastric furrow and, finding the centre, locate the gonopore. This is the only opening. Passing through the gonopore, the embolus will enter the uterus externus. This is a wide flattened pocket made out of transparent cuticle. The embolus will now explore the ventral wall of this pocket locating one of the paired openings to the spermathecae. (In fused spermathecae there is only one opening.) Once inside the embolus will now deliver the sperm where it will remain until egg-laying time. It will

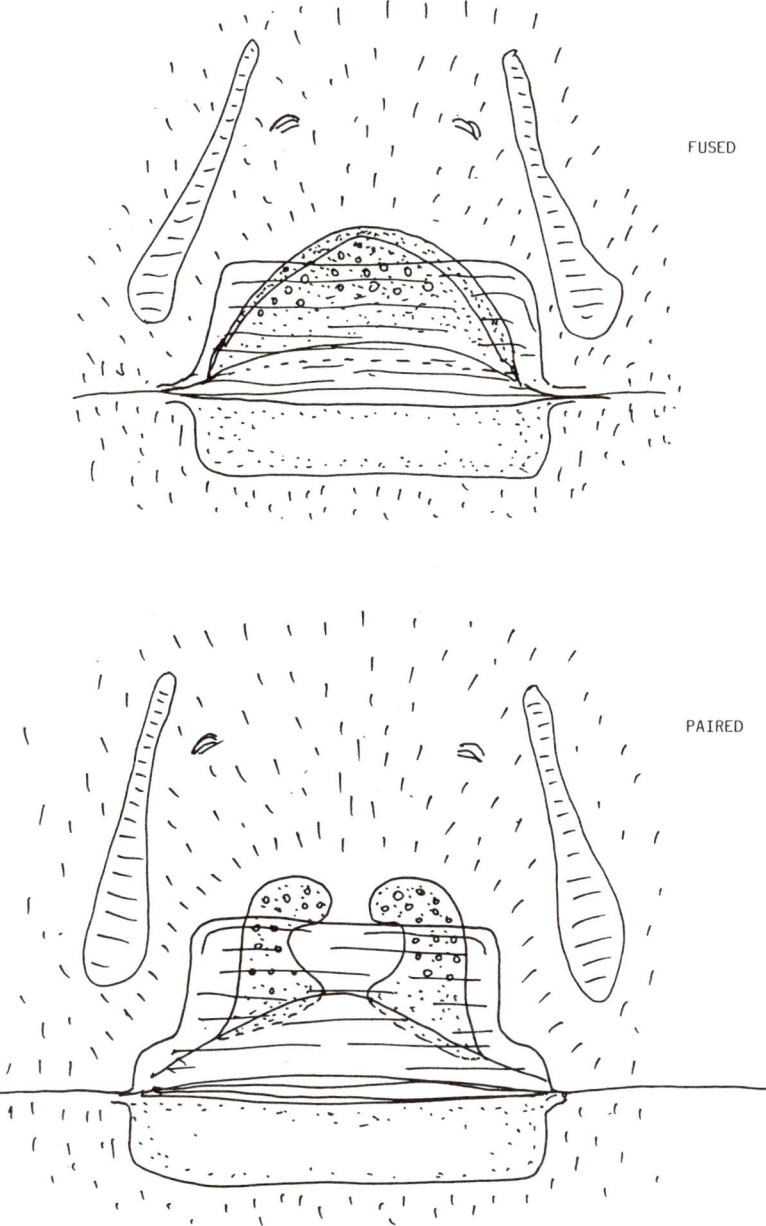

Fig. 31 The two basic forms of spermathecae.

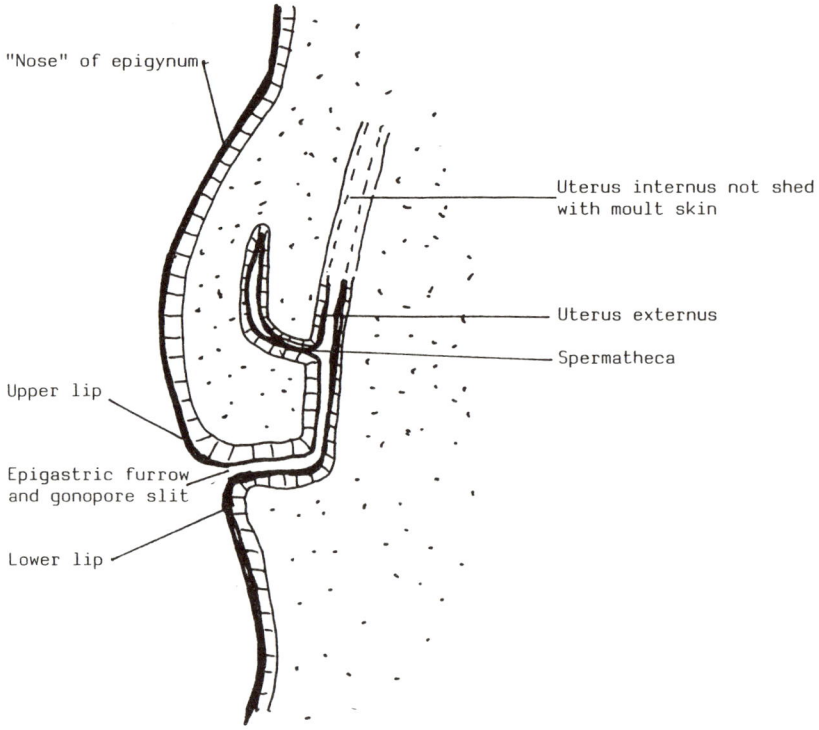

Fig. 32 Side view of the female's epigynum area.

then mix with and fertilize the eggs in the uterus externus as they are layed.

A female in most species should be quite obvious. It needs some experience to determine a male correctly. At first sight the whole epigastric area appears very hairy without a clear hairless furrow. The gonopore is very small compared to that of the female, and is surrounded by thick lips tightly pressed together. In some species a pair of accessory glands can be seen at either side of the gonopore. They look rather like a pair of trees drawn by a child. Unfortunately these glands are often mistaken for spermathecae. (Fig. 33) A simple rule of thumb will distinguish glands from spermathecae. If the base of the structure is wider than the organ itself, it is a spermatheca. If the base is narrower than the organ, it is a male accessory gland. Often in a male a mass of dark hairs can be seen behind the gonopore lips.

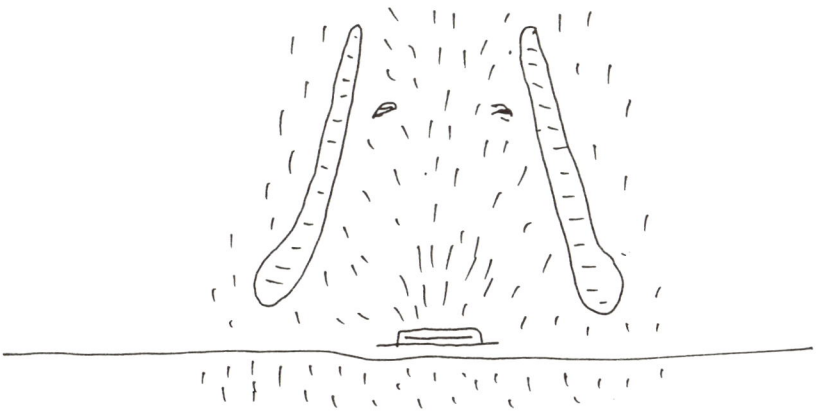

Fig. 33 Male epigynum area without and with accessory organs.

Sex determination of spiderlings is a very important area which offers great potential to a breeding programme. Males can be labelled and isolated for future use. Sex determination of moult skins as young as four months is possible. Higher magnification is needed for these small skins, at least X50. After a year old, sexing is relatively easy, giving positive results.

Chapter Eight

# Breeding

For many years tarantulas have been imported in large numbers into Britain and the rest of Europe. These creatures are captured in the wild and transported in far from ideal conditions. The result has been a drastic reduction in the wild populations of many species and many of the imported spiders did not survive the journey. Some died from parasites or disease. Many perished through ignorance of how to care for them. Many were males as these wander more in search of females so are more easily caught.

Breeding tarantulas in captivity overcomes many of these problems, not least preserving the species in the wild. Breeding can be a lengthy process particularly if a programme is worked out using sexed juveniles. Using imported tarantulas is unreliable as their age is not known, neither is their sex. However a mature male may be available and if the female appears in good condition mating can be undertaken.

Ideally the male should be recently matured, within the previous couple of months. His lifespan is very limited and in the short period between attaining sexual maturity and death his sole purpose is to find a female and mate with her. He will leave the burrow or web in which he has spent most of his life and go searching for the female, hence they are more easily captured. The female must not only be mature but be in breeding condition having recently moulted. It is therefore necessary to ensure that both male and female are ready to mate at the same time.

Assuming that this is the case they must be allowed time to get used to each other. It is dangerous to introduce one to the other without careful preparations. They will attack, believing the intruder is a predator. They should be allowed time to become aware that the opposite sex is nearby and this can be done by placing the containers side by side. They will drum to each other using their pedipalps and legs and this is often loud enough to be heard by the human ear. If he has not already done so, the proximity of a female will spur the male on to make a sperm web. It is only after this has been seen to be done that the next step should be taken.

Before mating the male must charge his palps with sperm. He constructs a web, generally at an angle of 45°, attached to the side of his container and the substrate. He moves underneath this, turns onto his back and deposits the sperm on the under side of it from his epigyne. He then moves above it and reaching underneath, sucks the sperm into his palps. This process can take a number of hours to complete. When he has finished the web is destroyed and the only evidence remaining is a fine, very white thread. If a mature male is seen to be in a peculiar position he is not moulting, but preparing to mate.

The female must be recently moulted and the ideal time to introduce the male is about six weeks after the moult. It is a waste of time trying to mate a female which is due to moult. She may go through the motions but when she sheds her skin the sperm in her body will be shed too. She becomes a virgin again and there is no chance that a mating prior to moulting will result in an egg-sac. It is important that the dates of moulting are kept accurately to avoid this situation arising. A female that is not in breeding condition is quite likely to attack the male.

Assuming the male has made his sperm web, the female is in condition and they are drumming to each other, the male can be introduced to the female. As an extra precaution he can be placed in a small container within the female's housing for a couple of days. It is vital that the male 'finds' the female as this is what happens in the wild. The female remains in her burrow or web and the male goes searching for her. The male can be released to approach the female, preferably after dark in subdued lighting. If all is going well the spiders will tentatively approach each other, touching legs. Sometimes this may take a couple of hours, but they cannot be hurried. When they decide to mate, the male will lift the female up using his front legs so that the mating hooks engage her fangs. He will then insert his palps in turn into her epigyne. A sign of a good mating is if the male jerks the female towards him using the still-engaged palp. This action helps to discharge the sperm which will be stored in the female's spermathecae until she is ready to lay her eggs.

Once mating has been completed the male will hurriedly retreat. It is at this point that the female may attack, so the male should be removed immediately. They can be kept apart by placing a sheet of plastic between them. It is advisable to repeat the process at a later date when the male has been seen to construct another sperm web. Mating tarantulas can be a rather nerve-wracking process for the owner. Some species, normally docile, can be very aggressive towards the opposite sex, generally the female rejecting the advances of the male. Aggressive species may be extremely gentle towards each other, for example the South American Bird eaters (*Pamphoboetus sp.*) and some species of Baboon spider. The Golden Star-burst Baboon (*Pterinochilus murinus*) which is very aggressive towards humans will actually share a container with a male and they will live together quite happily for a while. To be sure mating has occurred the spiders should be observed.

A male can be used to mate with a number of females, but the process of introduction will have to be repeated and the male be seen to construct a sperm web before each mating. His potency will gradually diminish as he gets older. Although a male Chilean Rose (*Grammostola spatulatus*) lived for two years after maturing, he was not fertile after about nine months. Males of species such as the Zebra (*Aphonopelma seemanni*) and Thailand Black (*Haplopelma minax*) have a life expectancy of only a few months after maturing.

Mating can be a very frustrating process for the tarantula keeper. Some species mate readily, whereas others are very reluctant. It is disappointing to have a pair of tarantulas apparently in good condition which want nothing to do with each other. Much more data is needed on how these species behave in the wild and the exact conditions they require.

Once mating has taken place the female should be placed in a clean environment. It is important that there are no mites present when she lays her eggs because these can destroy an egg-sac. Her container should be cleaned frequently until she is ready to make the egg-sac. The interval between mating and egg laying varies. It depends largely on the stage of development of the eggs within the female's ovaries at mating and could be up to seven months. An unmated female will have developing eggs in her ovaries. These may either be reabsorbed if she is not mated, or she may produce an infertile egg-sac. She will usually eat this when it does not develop which avoids wasting food.

When she is ready to make the egg-sac, she will first construct a saucer shaped web on which she will deposit the eggs. As they are layed the stored sperm will fertilize them. She will then wrap the eggs in the web to form a ball. It is at this point that mites can be enclosed with the eggs and destroy them. Baboon spiders such as the Feather Leg Baboon (*Stromatopelma grisepes*) build the egg-sac into the web, but in general most Tarantulas will carry it around until hatching is due. A female with an egg-sac will be very possessive of it and a usually docile spider may be aggressive. As she holds the egg-sac she will turn it frequently to prevent damage to the eggs at the bottom.

If a female discards her egg-sac and appears to have no interest in it, it could be infertile. They do sometimes set the egg-sac down, perhaps when feeding, but if she is disturbed she will soon pick it up again. If she has rejected it, it can be removed and opened very carefully. Infertile eggs will have the appearance of those freshly layed which are yellow. If they are fertile the beginnings of the spiderlings will be seen. If this is the case, all is not lost. The sac should be closed again and kept at a constant temperature of 26°C (78°F), and turned at least three times a day. Tiny spiderlings will emerge through the hole which can be enlarged at this time. They cannot open the egg-sac themselves, so if the female has not opened it six to seven weeks after laying it, it should be taken from her and opened. Most tarantulas are good 'mothers' and will care for a fertile egg-sac. After about six weeks the female will open the egg-sac herself and the tiny spiderlings will emerge, having moulted once inside it. The skins, pink in colour, can be seen. These tiny spiderlings will not eat for a while, but survive on the yolk in their bodies. It is quite safe to leave them together with the female until they moult again and begin to disperse around the container. After this moult, their second, they will be ready to feed and if they are not given suitably sized food they will begin to eat each other. They should now be removed from the female's container and housed in separate containers.

The number of young emerging from an egg-sac varies considerably. *Theraphosa leblondi*, the largest spider, produces about 60 whereas *Euathlus albopilosa*, the Curly Hair produces about 700. Some Baboon spiders produce more than one egg-sac. When one hatches another is layed, so the females have the ability to store the male's sperm for quite a long time and release it a little at a time. New World and Asian species will produce only one egg-sac between moults and mating again should be avoided until after moulting.

There have been occasions when hybrids have been produced. The male and female were of the same genus, but different species. There seems little point in doing this as it is against the interests of conservation and may eventually result in pure species being lost.

Breeding tarantulas can be interesting and exciting, but the resulting spiderlings must be considered. Mating must be undertaken in a responsible manner. A common species bred too frequently within a short period will result in many unwanted spiderlings. Captive breeding of rarer species will help towards their conservation. It may even happen that a tarantula is gravid when imported, having mated in the wild before capture. This will be an unexpected bonus.

Chapter Nine

# Rearing Spiderlings

Due to the increase in captive breeding of tarantulas spiderlings are often offered for sale. There are many advantages in rearing the young, but they are much more delicate than adults and are more exacting in their requirements. The following can be applied to spiderlings produced by one's own tarantula once they have left the egg-sac or to ones acquired elsewhere.

Once spiderlings begin to feed after their second moult (the first being inside the egg-sac) they must be housed separately to prevent cannibalism. This can be a problem with the more prolific species as the cost of containers can amount to quite a lot. The ideal size of container for ground dwelling species is 5cm. high by 5cm. in diameter and half filled with substrate. Too large a container will make it difficult for the spiderling to feed. Arboreal species should be housed in tall containers, ideally 10cm. high by 5cm. in diameter. They need a piece of bark on which to construct a web and about 1cm. of substrate to regulate humidity.

Young spiders can move remarkably fast so a method of feeding should be devised whereby they cannot escape when being fed or watered. A hole can be made in the lid which is then covered with Micropore tape, available from chemists. A small piece of tape is fixed to the main piece to prevent the spiderling sticking to it where it covers the hole. The tape can be pulled back and the food inserted through the hole. Alternatively, if a pooter is used for feeding, a cross can be cut in the lid using a sharp knife. The end of the pooter can be pushed through the hole and this will close again when it is removed. Both these methods allow air to move into the container, but extra holes can be made in the lid using a fine tool or in the rigid plastic using a fine drill. The lid should be completely removed once a week for ventilation. (Fig. 34)

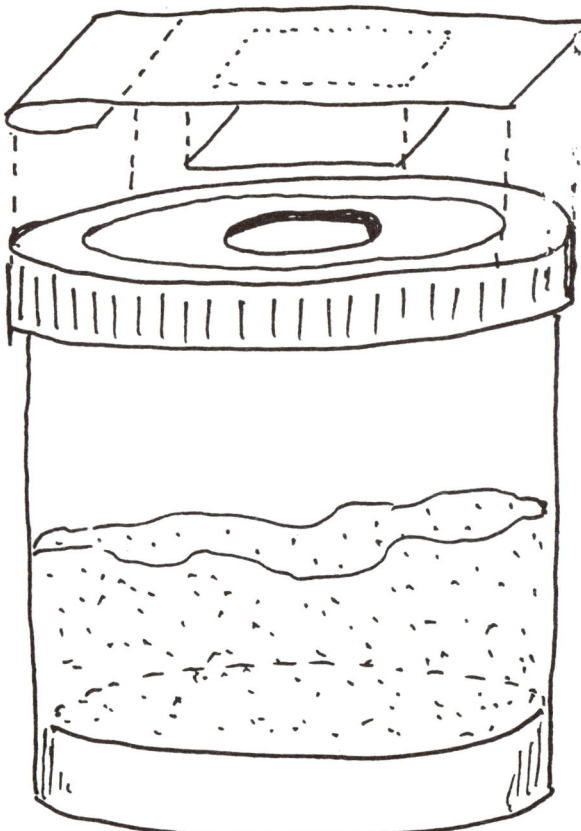

Fig. 34 Container to house a spiderling.

If a number of spiderlings are kept, the containers can be placed in either an aquarium or a cabinet to facilitate heating. The temperature should be kept at around 26°C (78°F). A small drop in temperature for a short period will not harm the spiderlings, but a rise may well kill them very quickly.

Well-fed spiderlings will moult at frequent intervals, perhaps once a month to begin with. Careful records of moulting should be kept to know when to avoid feeding. The skins can also be kept and, if set out and pinned, a picture of the spiderling's growth can be built up. These skins, after the fourth moult, can be used to establish the sex of the spider. (See 'Sex Determination'). The intervals between moults will become

A

B

# Plate 1

A: *Euathlus emilia*, The True Red Leg, a mature female.

B. *Euathlus smithi*, The Mexican Red Knee, a mature female.

A

B

C

# Plate 2

A: *Theraphosa leblondi*, The Goliath Bird Eater, side view, mature female.

B. *Psalmopoeus cambridgei*, The Trinidad Chevron, a sub-adult male.

C. *Aphonopelma seemanni*, The Costa Rican Zebra, a mature female.

A

B

# Plate 3

A. *Cyclosternum fasciatus*, The Tiger Rump Doppel-ganger, a mature female.

B. *Poecilotheria regalis*, The Indian Ornamental, a mature female showing the yellow colours under the legs.

## Plate 4

A. *Latrodectus mactans*, The Black Widow, a female showing the distinctive hour-glass marking. (Photo: Roy Dunn)

B. *Steatoda paykulliana*, The False Widow. (Photo: Roy Dunn)

C. *Araneus diadematus*, The Common Garden spider, a female in her web. (Photo: Roy Dunn)

D. *Nephila madagaskariensis*, The golden Silk orb web spider.

E. *Sicarius species*, A Six-eyed Crab spider from the Namid Desert which has venom more deadly than most.

F. *Heteropoda venatoria*, The Giant Huntsman spider having just opened her egg-sac. A spiderling is emerging. (Photo: Roy Dunn)

Fig. 35 Spiderling rearing containers.

progressively longer, until, when adult, the interval is a year or more. Feeding spiderlings need not be a problem with the correct size of food available. It is surprising how large an item of prey a young tarantula will tackle. They will often take food as large as themselves, particularly maggots. Micro crickets are the most commonly used food, but these must be removed if uneaten and moulting is imminent. Fruit flies will not cause any harm to the spiderling so can be used more safely. A pooter is invaluable when dealing with small food.

As the spiderling grows it must be moved to larger quarters. If it is kept in cramped conditions it will not be able to moult successfully. The container size should be increased in proportion to the size of the spider and it may need five or six containers before being large enough to move into its adult housing. The rate of growth varies considerably between species. A slow growth rate indicates a long life ahead.

Correct humidity is very important for small spiderlings which can dehydrate very quickly. For this reason the container should have as few air holes as possible. The majority of species require the substrate to be damp, but not so damp that a lot of condensation forms on the sides of the container. The spiderling can become trapped in the film of water and be unable to escape. It will drown. To be sure that this will not happen it is wise to experiment with the containers before the spiderlings are put in. Some species require a fairly dry atmosphere. *Ceratogyrus darlingi*, the East

African Horned Baboon, comes from an area with distinct wet and dry seasons. In the wild the spiderlings emerge from the egg-sac in the dry season. Until this was known many captive bred ones did not survive when kept in what was considered a suitably humid atmosphere. Much more information is needed on specific species and their natural habitat before all mistakes can be avoided, but reference to the climate of the country of origin is helpful.

Moving spiderlings from one container to another can be a problem as they move so fast. It is quite safe to leave them in a cool place such as an unheated room for a few minutes before attempting it. This will slow them down. As an extra precaution place both containers in the bath (with the plug in) and encourage the spiderling to walk into the new one. If it does try to escape it can easily be caught.

Rearing spiderlings can be very satisfying particularly if breeding is to be carried out in the future. It makes no demands on the wild stock and should eventually lead to a reduction in imports. Buying spiderlings is an inexpensive way to build up a collection and the tarantula will eventually have its entire adult live ahead of it and be free of disease or parasites.

Chapter Ten

# Atypidae, Dipluridae and Ctenizidae

The Order Araneae is divided into three Sub-orders, Mesothelae, Mygalomorphae and Araneomorphae. Mygalomorphae, the Sub-order which includes Theraphosidae, the tarantulas, also contains Atypidae, the Purse Web spiders, Dipluridae, the Funnel Web spiders and Ctenizidae, the Trap Door spiders. They all have downward striking fangs and two pairs of book-lungs, the same as tarantulas. Atypidae has been included because in the Sub-order Mygalomorphae it is the only one to be found in Great Britain. The species is *Atypus affinis*. It has been recorded throughout the British Isles but is most common in the southern counties where they are often found in colonies. The female has a body length of 12mm. She has a light brown abdomen and the cephalothorax and legs have a greenish tinge. The male is 8 to 9mm. and is very dark brown, almost black. The chelicerae are very large with teeth on the underside which are used to perforate the wall of the tube web. (Fig. 36)

The common name of Atypidae is Purse Web spider. They spend their whole life sealed in a silken tube. This tube is up to 30cm. in length and 1cm. in diameter. Most of this is buried in the ground, but a short finger-like extension protrudes horizontally, lying on the surface. In years past these spiders were a puzzle. It was not understood how they fed. The theory was that they lived on earthworms which strayed into the burrow because the tube had no visible entrance or exit. The extension is the spiders hunting ground. It is well camouflaged with leaves, twigs and earth, and when an insect strays onto it the spider strikes. It spears the prey through the silken wall with its sharp curved fangs. It then uses the teeth on the basal segment of the chelicerae to cut a hole through the silk. The prey is then pulled into the burrow and taken to the slightly wider chamber at the base. Before the spider eats it repairs the hole in the tube using its fangs to hold the edges together while it mends it with silk from the spinnerets. The female will never leave her burrow, so the male must

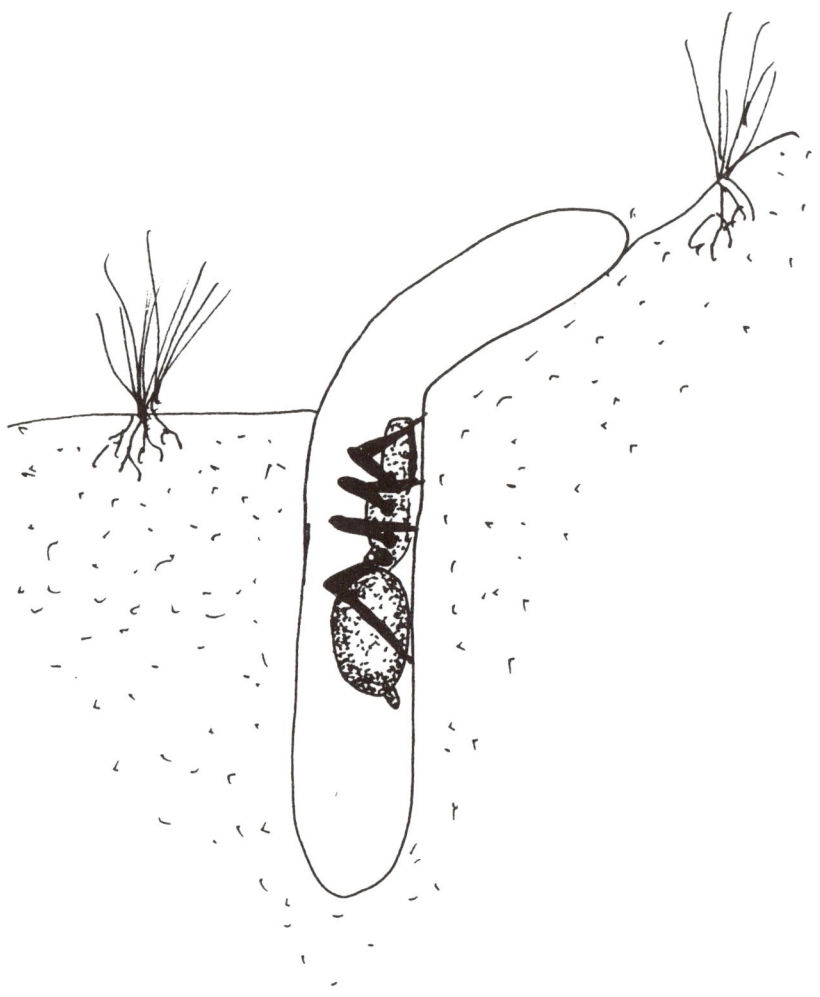

Fig. 36 Purse Web spider in its tube.

go in search for her. After his final moult in autumn he will emerge in his adult colours. When he finds a tube of a female he approaches cautiously then taps rapidly with his palps and legs. He repeats this drumming and if there is no response from within he tears a hole and enters. Having given the correct signal he is safe from her fangs. If the female is unreceptive, either immature or gravid, she would signal to him, giving a warning tug on the tube. After entering the tube, he mates with her. He will not leave the tube again but they will live together for several months. It is not until the following summer that the eggs are layed, about 100 in an elongated silken sac hung from the wall of the lower chamber. In autumn the spiderlings emerge from the sac. For the next six months the mother cares for them in the tube. When spring arrives it is time for them to disperse. The female makes a hole in the tube and the young climb out and onto vegetation. From here they 'balloon' to new sites by trailing silk which is caught by the breeze. They then begin their own burrows and tubes. Maturity is not reached until the spiders are about four years old. It is remarkable that it takes about 18 months from mating to dispersal, meaning that a female can only reproduce in alternate years. The lifespan is thought to be 7 to 8 years.

The Purse Web spiders spend their entire adult life within the silken tube. Collecting should be kept to a minimum in Britain as these spiders are quite rare. Atypidae are found throughout the world so collecting can be done where they are more numerous. It is best to dig up the tube, complete with spider, without damaging it. Unfortunately these spiders are not easy to observe except when catching their prey.

Dipluridae are generally quite small spiders and are mostly black. Their distinguishing feature is the length of their spinnerets. They are profuse web builders and have proportionately longer spinnerets than other mygalomorphs. Dipluridae are one of the few mygalomorphs to use web in the capture of their prey. In Theraphosidae it is mainly for the construction and lining of retreats. There has been little work done on Dipluridae webs, some being so complex that there seems little form or pattern to them.

Many species live in the leaf litter beneath trees and bushes. Their webs, the top layer at least, are conspicuous. They look very much like the web of the House Spider, *Tegenaria domestica*. There is a funnel shaped retreat and a sheet of web in front of this. In Dipluridae there are three or four layers of web all inter-connected. In the wild the web is camouflaged with leaves and debris and really looks quite a mess. Species which have been kept in captivity do not have access to such debris and can be observed more closely.

The common name of the Dipluridae is Funnel Web. This is often considered synonymous with deadly species because of the reputation of the Sydney Funnel Web, *Atrax robustus*, which is a diplurid. The majority of species are relatively harmless.

The largest diplurid in Europe is *Macrothele calpeiana* which is found in southern Spain. It is a black spider with a dull abdomen but glossy black carapace and legs. A female grows to about 3cm. body length. When kept in a container 30cm. × 15cm. × 10cm. it fills it with layers of web with inter-connecting tunnels. It has a retreat at one end. Prey introduced becomes entangled in the web and the spider rushes out to immobilise it. It takes it back to the retreat. When a male is introduced he can be safely left with the female until the egg-sac is made. This she will carry around or suspend it from the web whilst hunting. It takes three to four years for a female to mature and they produce about 100 spiderlings at a time. *M. calpeiana* has a reputation for being dangerous in its native habitat but there is no evidence that its venom is particularly potent.

A small diplurid from Kenya, *Thelechoris karschi*, in captivity makes a similar web. In the wild their webs are in trees and bushes, where a number live together with other species in close proximity. The only way to collect these spiders is to take the whole mass of web into a sweep net. It is impossible to catch the spiders otherwise as they have so many escape routes. *Ischnothele guyanensis* is a similar species from Guyana.

Temperature and humidity in captivity depend on the country of origin. Dipluridae can be kept in quite a small space, but if given a larger container will soon fill it with web. These are interesting little spiders to keep as they are quite active.

Ctenizidae are the Trap Door spiders. This is another mygalomorph which spends its whole life in one place, enlarging its burrow as it grows. This is not only a place in which it can hide and provide protection from predators, but a place from which it can strike out at its prey.

Ctenizidae have a smooth, rather shiny appearance. On the chelicerae there is a rake of thick short setae which are used along with the fangs to excavate the burrow. The setae are known as the rastellum. The spider compresses the excavated soil into a pellet and throws it out of the open door. There are two types of door, 'wafer' and 'cork'. The 'wafer' door is quite thin, constructed of web and camouflaged with debris such as soil particles which are cemented together with saliva. The 'cork' door is thicker and has bevelled edges which fit the rim of the burrow exactly. It is constructed of the same materials. The door is added to periodically as the spider grows and enlarges the burrow. These doors are extremely difficult to spot as they match their surroundings perfectly – that of a long established spider may have moss growing on it. (Fig. 37)

The door is the first line of defense for trap door spiders. It will be unnoticed by many predators which may be scorpions, centipedes or parasitic wasps. One species has even adapted to hold its door shut if danger approaches. It has spines on the third pair of legs which it presses into the sides of the burrow while it holds the door tightly closed with its fangs. An Australian species, *Anidiops villosus*, has a second line of defense.

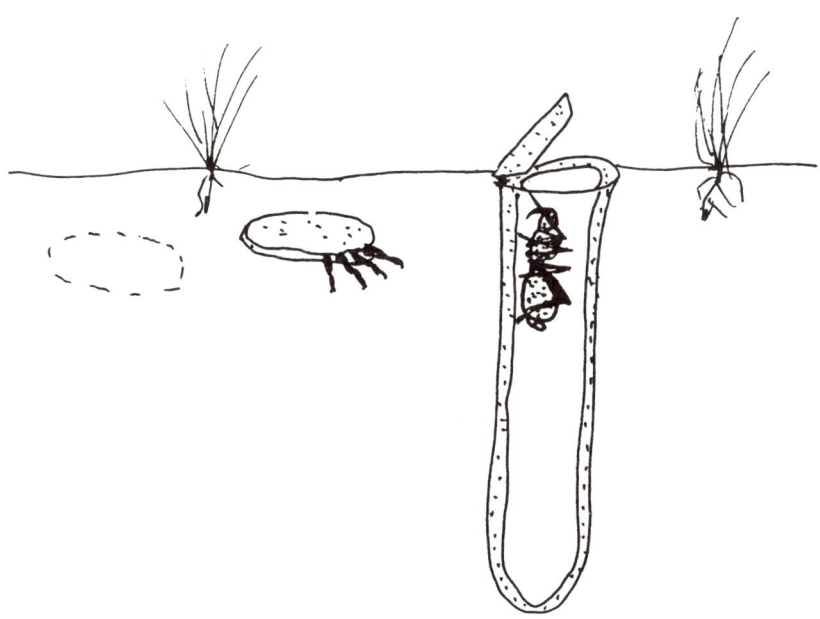

Fig. 37 Trap door spider in its burrow.

It stores its rubbish, remains of food, behind the silk linings of the walls of the burrow. If threatened, as it hurries down its burrow, it releases these remains in passing. The web to which they are stuck forms a false bottom to the tube and the predator assumes it is unoccupied and leaves. Other species have a side tunnel, also with a trap door. When threatened they retreat into this and close the door. The burrow then appears unoccupied. A Venezuelan species, *Rhytidicolus structor*, goes even further. It has three trap doors, so if the predator pursues it through two, it will generally be safe behind the third. Some species do not go to such lengths to avoid predators. They have a 'back door' through which they can escape. They excavate a side tunnel from the main one, and if a predator threatens they can escape through this, the entrance being disguised with leaves and earth. *Stothis astuta* from South America makes a U-shaped burrow with a trap door at each end. It can make its escape from either door.

One genus, *Cyclocosmia*, has not developed feats of engineering to protect itself. It has modified its abdomen. Species of this genus are found in the United States, Mexico and Indochina and they have a flattened posterior which fits the diameter of the burrow exactly at the bottom. The skin on this part of the abdomen has grooves radiating from the centre and a leathery texture. The predator, on entering the burrow, believes, hopefully, that it is empty, the rear end of the spider forming the bottom. Some parasitic wasps, however, can penetrate the tough skin. As well as protecting the spiders from predators, the trap doors can ensure that the correct level of humidity is maintained. They also protect the spider from flash floods.

Ctenizid spiders are rarely seen. Some open the trap door slightly and wait for passing prey. They do not like to leave the burrow and often the rear legs remain inside. It takes only a fraction of a second for the spider to strike and the prey is dragged inside where the spider consumes it at its leisure. It is probably the sensitive hairs on the legs which pick up vibrations. Some species extend their strike distance by laying twigs which radiate from the burrow.

When a male matures he will leave his burrow to search for a female. The male entices the female upwards. Many species mate at the entrance to the femlae's burrow, with her still not leaving it. The female cares for the egg-sac until it hatches and the spiderlings emerge. As with *Atypus affinis*, periods of adverse conditions, for example the dry season, will be passed within the burrow, with the spiderlings emerging when there is a plentiful supply of food.

Keeping trap door spiders in captivity can be rather disappointing. They may never be seen, the only indication that they are still alive being that the prey has disappeared. If in doubt as to whether the spider is still alive or not, the trap door can be held slightly ajar with a twig. If the spider is fit and healthy it will remove this and shut the door to regain its privacy. The species that have been available have not been very large and have not been bred successfully in captivity. If these spiders are collected, it is kinder to extract the entire burrow with the earth around it.

Chapter Eleven

# Araneomorphae – The True Spiders

Araneomorphae, or true spiders, are, in terms of evolution, more advanced than the Mygalomorphae which include tarantulas. They are the spiders which can be found anywhere from the suburban garden or town flat to the desolate tundra of the Arctic or the rain forests of Equatorial regions. They have evolved fangs which move on an axis parallel to their body which means they do not have to strike downwards against a stable surface. They have only one pair of book-lungs, the other pair having developed into trachea making respiration more efficient. Some have developed very unique ways of trapping their prey. On the whole they can be much more interesting than tarantulas although they never reach the same proportions as many Theraphosidae. Their lifespan is much shorter, therefore their metabolism is faster and they grow more quickly. Many are also more active.

Exotic species of true spider have been available to the enthusiast for some time. The most popular species would appear to be the Widows (*Latrodectus species*). All members of this genus are considered dangerous and in the United Kingdom a Dangerous Wild Animals' licence is required to keep them. This licence is issued by the local authority which has to be satisfied that every precaution is taken to prevent escapes. The cost of this licence in some areas, plus the compulsory insurance, makes it unrealistic to keep dangerous species, except for scientific study. There have been few fatalities, but the law should be observed.

Wolf spiders of the Family Lycosidae are popular. These are hunting spiders and include *Lycosa tarentula*, the true tarantula. These are interesting to keep because they are so active. Orb web weavers, such as the giant *Nephila* species from the tropics are impressive, but imported specimens usually have only a short life left. Other smaller orb weavers of the Family Arianeidae, although not as impressive, are interesting. Fast moving members of the Family Sparassidae, known as Giant Huntsman spiders or Banana spiders, are easy to keep. Wandering spiders of the Family Ctenidae are sometimes available, but *Phoneutria fera* from Brazil should be avoided as it possesses one of the most toxic of all spider venoms. There are many smaller species of spider which can be kept successfully in captivity, both from tropical and temperate regions. As long as the conditions in which they are kept are appropriate to their lifestyle and climate of origin, they will live and reproduce quite happily.

A question often asked is 'How venomous are spiders?' Certain species have the reputation of being deadly, but these are by far in the minority. The most notorious must be the Black Widow (*Latrodectus mactans*), but the incidence of fatalities from bites is very rare. The purpose of the spiders' venom is to subdue and kill then help in the digestion of their prey. As they do not eat people, they only bite when inadvertently touched or frightened. Unless cornered they will try to escape. Anyway, the majority of spider species do not have fangs strong enough to puncture human skin.

The poison produced by spiders is a protein which varies considerably from one species to another. The venom of *Latrodectus species* is neurotoxic, it effects the victim's nervous system. Symptoms occur within 20 minutes and include excruciating cramp-like pains in the chest and abdomen. Other symptoms may be difficulty in breathing, blurred vision and perspiration. Although the symptoms are dramatic there is little threat to life in a healthy adult. There is an anti-venom available.

The venom of the Violin spiders (*Loxosceles sp.*) is cytotoxic, it destroys cells. The bite itself is not painful and the first signs will be a pimple or small sore. About two hours later pain can be felt and within a day or two the sore ulcerates with the surrounding tissue becoming dark, swollen and inflamed. The area becomes intensely painful. After about two weeks the sore dries, as long as secondary infection has not occurred, and a scab forms. When this eventually falls off, heavy scarring will remain. There is no anti-venom available and although the bite is not life-threatening, it can have serious cosmetic effects where necrosis has occurred.

By far the most dangerous are the Six-eyed Crab spiders (*Sicarius species*). They should not be confused with Crab spiders of the Family Thomisidae which pose no threat to humans. (Beware common names!) *Sicarius* species are ground dwelling generally in arid regions of the world. They are flattish spiders which spend their time buried just under the surface of the sand for extremely long periods, even months, waiting for prey. These spiders are not aggressive and are usually reluctant to bite. They prefer to 'play dead' so should be left well alone. Their venom is extremely toxic being made up of a 'cocktail' of cytotoxic, neurotoxic and haemotoxic poisons. It can cause an intensely necrotic haemorrhagic lesion at the bite site, internal haemorrhages, paralysis and eventually death. There is no anti-venom. The prognosis for victims is not good and painful death would occur, but because of the secretive nature and non aggression of these spiders, bites are rare.

Spider venom is beginning to play its part in medical research. Venom from certain species is being used to develop drugs for the treatment of stroke victims and epileptics. If spiders are treated with respect, they will not attack. Their bite is a last resort if cornered and few are fatal unless the victim is a small child or has another medical condition. The easy answer is not to get into a position where one can be bitten.

**Family Theridiidae**. Members of this Family are known as Scaffold web spiders, due to the construction of the web. (Fig. 38) They will, unless disturbed, spend their whole lives on this web, so in captivity only require a small living space. Once established these spiders are reluctant to leave their web, but precautions should be taken that they cannot escape. They prefer flying insects as food, although they will feed on appropriately sized crickets and locusts. It is interesting to watch them feed. They wrap their prey in silk and, unlike tarantulas which 'chew' their food, they inject digestive juices into the fly rather than onto it, then suck the resulting mixture out of it leaving an empty shell. They breed readily, producing a number of egg-sacs which are suspended from the web. When the spiderlings emerge the female helps them to feed by killing the prey then puncturing it many times. The young gather round to feed. Problems can arise with breeding as the spiderlings are very small. Great care should be taken that they do not escape especially if they are of a species of *Latrodectus*. Despite their name Widow, they do not automatically kill the male after mating. In fact they are quite tolerant of him and a pair will live together for a long period.

Not all members of the family are considered dangerous, although all have a nasty bite. The Widows can be recognised as they have a characteristic marking on the ventral side of the abdomen. It is shaped like an hour-glass and red in colour. Apart from *Latrodectus mactans* from the United States of America, there is *L. hesperus*, the infamous Australian Red Back. *L. geometricus*, the Brown Widow is found in Africa and around the Mediterranean. There are other species of *Latrodectus*, but these three are the most commonly kept. They all require a licence in Britain.

*Steatoda paykulliana*, the False Widow, is also a member of the Family Theridiidae. It is very similar to *Latrodectus* but does not require a licence. It is black with red markings on its dorsal side. It does not have the hour-glass on the ventral side. It has all the features of the Widows without the potent venom. It makes a similar web and helps the young feed. The False Widow is an excellent substitute for the true Widow.

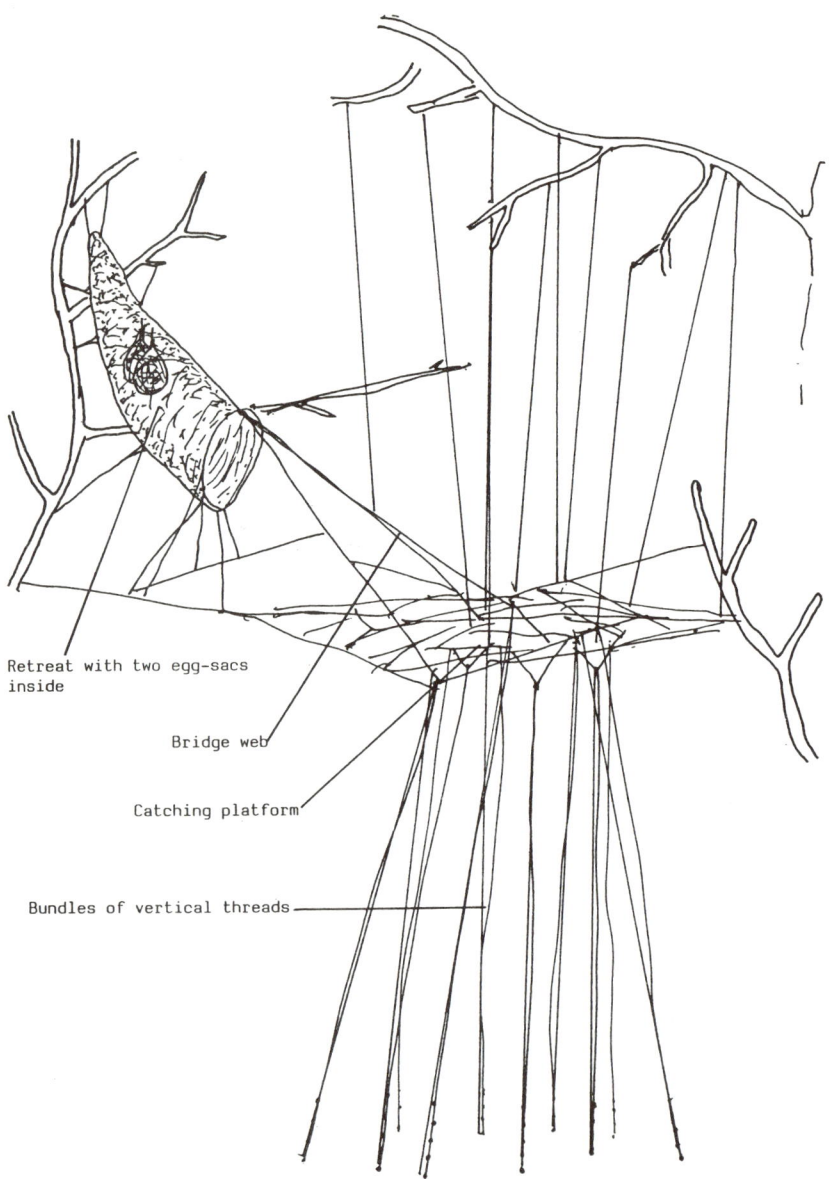

Fig. 38 Scaffold web.

**Family Lycosidae**. Members of the Family Lycosidae are the Wolf spiders, known as such because they are hunters. The true tarantula is the most popular member of his Family. Wolf spiders do not produce web for trapping food. They rely on speed. As they are not sedentary and do not have a web to which they can attach the egg-sac, the female carries it around. It is attached by silk to her spinnerets and she will continue to guard it in this manner until it is ready to hatch. The egg-sac splits and as the young emerge they climb onto the mother's back. In this way she continues to keep a watchful eye on her offspring until they moult and become independent and begin to feed. The most common member of the Family Lycosidae in the United Kingdom is *Pardosa amentata* and these are often seen during the summer months in gardens, carrying their egg-sacs or young. This species is less than 1cm. long, but there are much more impressive species from warmer climates, up to 4cm. in length. These often dig burrows, particularly where there is a risk of very low humidity or extreme heat in the open. These burrows can be as much as 20cm. deep.

Due to their life-style these spiders need a fairly large area, about 25cm. square. The floor should be covered with sand or peat and a shelter provided. Consider the country of origin of the species being kept – was it from the desert of Arizona or an area of Africa with distinct wet and dry seasons? It may even be a European species which does not require extra heating. Knowing the country of origin is very important as members of this Family come from a variety of habitats. Food which will settle on the floor is more suitable than, for example, flies which tend to settle at the highest point in the container. Even if it is a desert species, water should be available for drinking.

An interesting feature of some species of Lycosidae is their courtship. Some drum with their palps, just as Theraphosidae do, but more usual is that the male signals to the female in a type of semaphore, using his palps and forelegs. This signals his intentions to her.

**Family Pisauridae**. Closely related to the Family Lycosidae is the Family Pisauridae. This includes *Dolemedes sp.*, the Raft spiders. Two species, *D. fimbriatus* and *D. plantarius* are found in Britain, but they are more plentiful in Europe. *D. plantarius* is on the Red List (Endangered species) in Britain, so it should not be taken from the wild. *D. fimbriatus* is one of the largest British spiders and is found near water, swamps and marshes. The adults use the water in the same way that other spiders use their web, to feel vibrations. They place their forelegs on the surface. They can take quite large prey, sometimes even small fish, hence their alternative name, the Fishing spider. They can be kept in a small aquarium with about 1cm. water covering half the area, the rest being moss and twigs. When mated the female carries the greenish egg-sac around until it hatches. In the wild the spiderlings disperse and usually remain on tallish plants until adult. This is an attractive spider to keep.

Also belonging to the Family Pisauridae is *Pisaura mirabilis*, the Nursery web spider. Courtship is fascinating in that the male kills and wraps a fly in silk and presents it to the female before mating. When the female lays her eggs she carries the cocoon around under her sternum. When ready to hatch, she builds a tent-like web in vegetation which forms the nursery. She guards this until the spiderlings disperse.

Although some of these species may not be impressive in size and may be regarded as 'common European spiders', they warrant close observation. They do not require a lot of space, neither do they need heating. Some are easily obtainable – from the English countryside. There are equivalent species in temperate latitudes around the world, but tropical species are more appealing due to their size.

**Family Nephilinae**. Of the exotic spiders available, members of the Family Nephilinae are surely the most impressive. In the wild they construct huge orb webs up to two metres across of golden silk. This silk is immensely strong and the web can stay in situ throughout the spider's life, being repaired when necessary. This is one of the main problems with imported adults of *Nephila sp*. They have to construct a new web from scratch so few survive. An adult female can have a leg span of up to 17cm. with a body length of 5 to 6cm., so they are virtually helpless without their web. As these spiders can be quite expensive it is best to try to obtain captive bred juveniles and allow them to become established where they will not be disturbed. There are a number of captive bred species available including *N. senegalensis* and *N. madagaskariensis* which came to the United Kingdom as egg-sacs and the spiderlings have been successfully reared and bred. Unless one is lucky enough to obtain a wild caught gravid female which survives long enough to produce an egg-sac, it is virtually impossible to breed from wild stock. The simple reason is that males are not imported. Collectors look for large specimens and overlook the minute males. These are so tiny it is difficult to believe that they are of the same species. Some are only one thousandth of the size of the female. Being so small they are overlooked as items of food by the female so can mate with her in complete safety. They usually take up residence in her web long before she is mature. He waits for her final moult, then quickly moves in to mate with her.

*Nephila* do not require a container. They will happily construct their web in the corner of a room and not leave it. As long as the room is warm enough and the web is sprayed occasionally the spider will survive. It can be fed by throwing prey into the web. If a female is mated and produces an egg-sac, this will be attached to a surface at the top of the web. The spiderlings can be collected when they begin to disperse a couple of weeks after emerging. Alternatively a frame can be made, about a metre square with netting on both sides and suspended from the ceiling. This should be sprayed to allow the spider to drink. *Nephila* do not require cleaning except to remove remains from the bottom of the frame. The web should not be disturbed as they will be helpless without it. (Fig. 40)

Fig. 39 *Nephila senegalensis.*

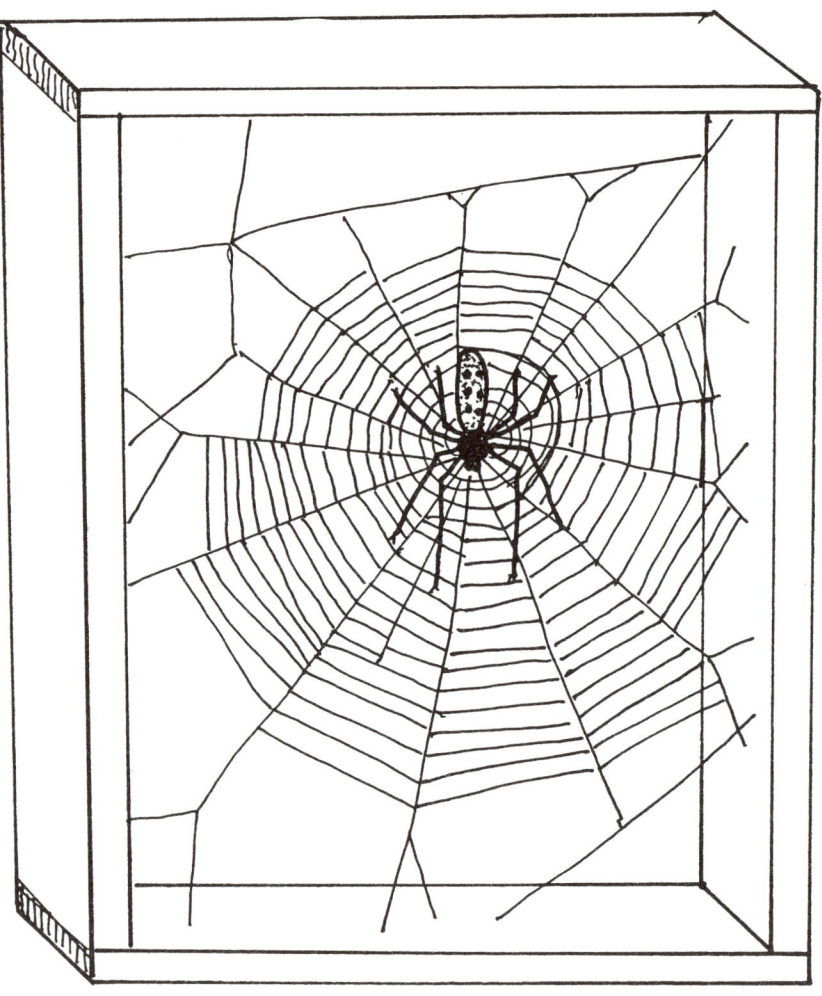

Fig. 40 Nephila in a frame.

**Family Araneidae.** This is another Family of orb web weavers. They are less impressive than the *Nephila*, but are easier to keep. Argiopes are found in the tropics as well as in more temperate latitudes. An attractive species from Europe is *Argiope bruennichi*. This is an annual spider. It overwinters as eggs with the spiderlings emerging in the spring. By autumn it has mated and the egg-sacs are produced. Once this is done the spider dies. *A. bruennichi* has a black, yellow and white striped abdomen. It constructs its web, up to 30cm. in diameter, in grass and low vegetation. A feature of this web is a zig-zag stabilimentum, thick strands of silk, vertically through the centre. (Fig. 41) The egg-sac is shaped like a vase, the top of which opens when the temperature rises in the spring. (Fig. 42) A female has a body length of 4cm. whereas the male is less than 1cm. These spiders can be kept in a frame about 30cm. square, covered on both sides with netting, as for Nephila. Temperature is dependent on the country of origin of the species, but *A. bruennichi* can be kept without heating. Flying prey is preferred.

A spider which can be observed in Britain without removing it from its habitat is *Araneus diadematus*, the common garden spider. These make an orb web up to 40cm. in diameter and are most noticeable from late summer to autumn. The females produce their eggs in autumn, but before this have enormous abdomens. The pattern varies, but is often in the form of a cross, hence another common name, the Garden Cross spider. Like the Argiopes, this is an annual spider. When its web becomes damaged it can completely rebuild it, but before doing so it eats the old silk to avoid wasting precious protein.

Orb weavers sometimes brought from tropical areas are *Gasteracantha* species. Although small, these spiders have amazingly shaped abdomens which are brightly coloured. They will require temperatures of around 26°C (78°F).

**Family Sparassidae and Family Ctenidae**. A common import are the Giant Huntsman spiders of the Family Sparassidae. These are usually *Heteropoda venatoria*, a very fast moving araneomorph. They have arrived in Britain in bunches of bananas, so have been called Banana spiders. They are cosmopolitan and are found on every continent in tropical regions, probably because they are such good travellers, hitching lifts on ships. They are somewhat flattened in appearance with legs which all seem to point forwards. When at rest they can slip under loose bark or stones or flatten themselves against tree trunks. They have adapted well to human habitations in their native lands and are welcomed as they control the populations of cockroaches and other unwanted insects. They live well in captivity and have been bred successfully. They prefer to rest on a vertical surface so a tall container is best. To minimise escapes the container can be inverted so the lid is at the bottom and can be removed safely. These spiders tend to move upwards if disturbed. A piece of bark attached to the side allows the spider a retreat. A female has a body length of about 2.5cm. and the male is slightly smaller. When the female produces an egg-sac, it is a pinkish disc which she carries around until the spiderlings emerge. As these are tropical spiders they require a temperature of around 23°C (75°F). Care should be taken not to make the container too humid, especially if there are spiderlings, as they can

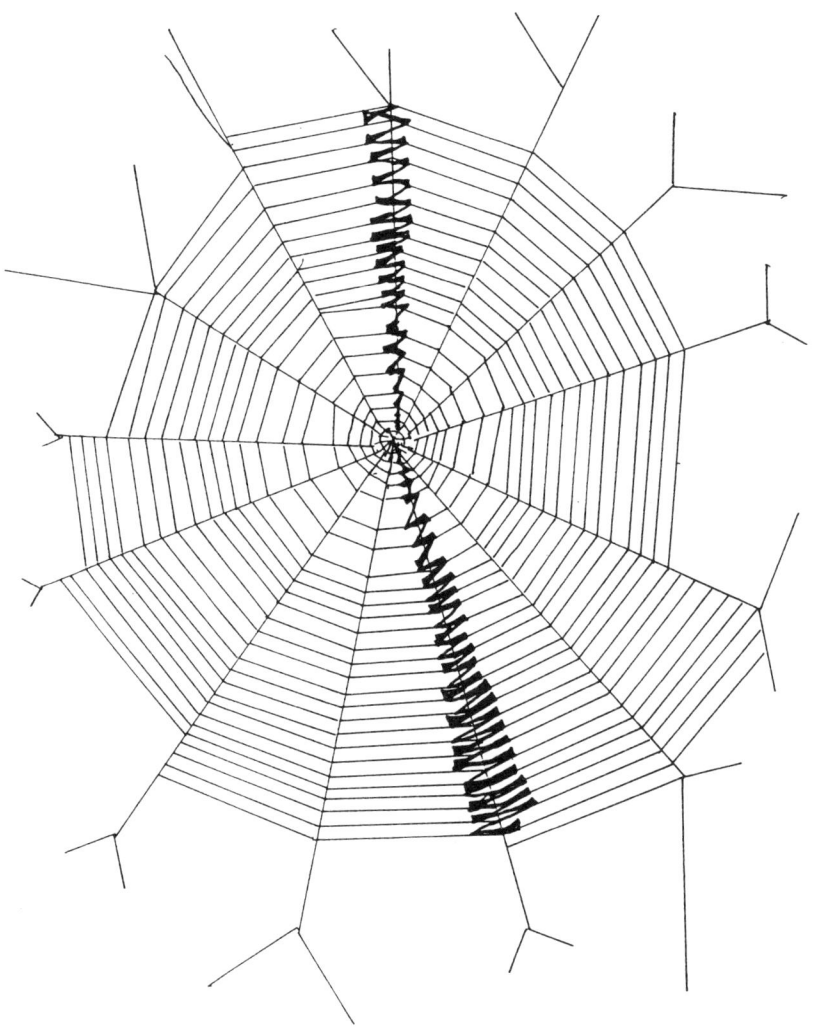

Fig. 41 Web of *Argiope bruennichi* showing stabilimentum.

become trapped in the condensation. There are many interesting species of Sparassidae, but *H. venatoria* is the most common.

Wandering spiders of the Family Ctenidae have a similar body shape and stance to the Sparassidae. *Cupiennius salei* is an attractive species from Panama, having a body length of about 3cm. It is marked in black and

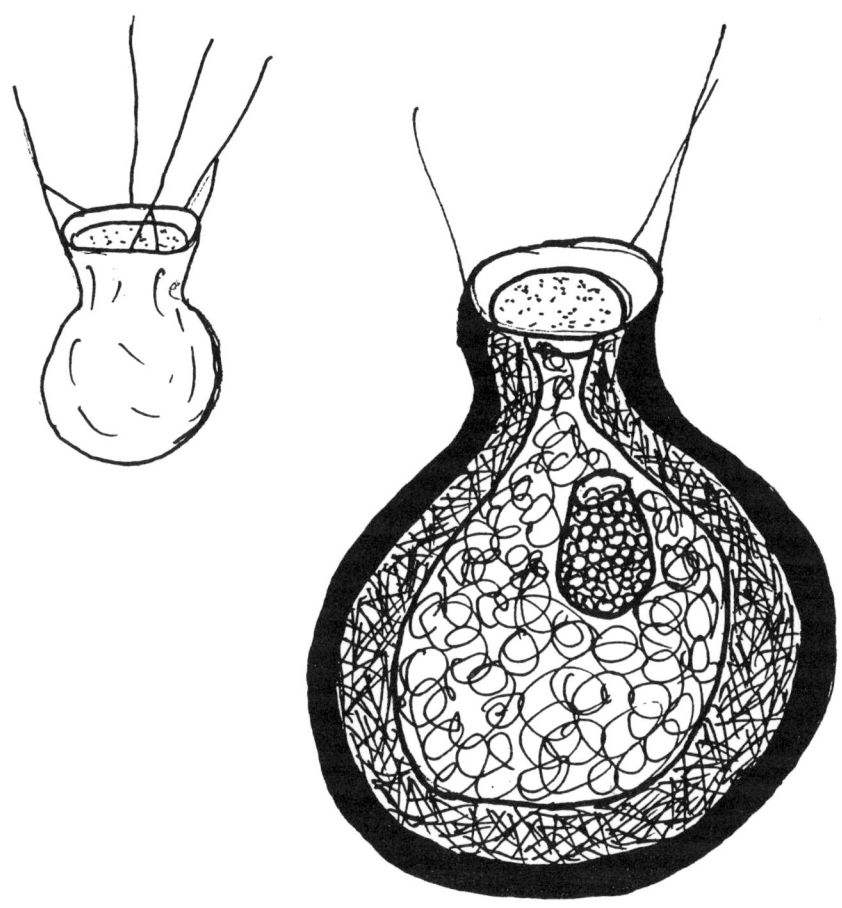

Fig. 42 Egg-sac of *Argiope bruennichi*.

orange/brown. Ctenidae, like Lycosidae and Sparassidae, carry their egg-sacs. When breeding them in captivity a sign of an infertile egg-sac is when the female rejects it. *C. salei* can be kept in a similar container to Lycosidae and the temperature should be between 20°C and 23°C (70°F – 75°F). One member of the Family Ctenidae is considered dangerous, so requires a licence in Britain. It is *Phoneutri fera* from Brazil and has one of the most potent of spider venoms.

Any species of araneomorph can be kept and bred successfully in captivity, given the right conditions. There are thousands of species to choose from and they do not need to be large to be interesting. The

Family Salticidae, the Jumping spiders, are fascinating to watch, from the common Zebra Jumping spider, *Salticus scenicus* on house walls, to *Portia africanus*, a tropical Salticid which looks like a dead leaf and preys on other spiders. *Scytodes thoracica*, the Spitting spider traps its prey by spraying it with a substance composed of glue and poison which pins it down, just as Gulliver was tied down. The Daddy-long-legs spiders, Pholcidae, are expert at wrapping their prey using their long legs. *Tegenaria domestica* or *T.gigantea*, both found in British houses and sheds, make a sheet web with a retreat at the end of it. If food is placed on the web, the spider rushes out to capture it.

Each Family of Araneomorphae can provide interesting members, no matter how humble or common. They are well worth a second look.

Chapter Twelve

# Scorpions

Scorpions can be roughly divided into two groups, those which have a potent venom likely to cause death, or at least serious illness, and those whose venom will cause similar effects as a bee or wasp sting. The former group belong to the Family Buthidae and to keep any of these in the United Kingdom requires a Dangerous Wild Animals' Licence. There are many species of scorpion that are interesting to keep and are not potentially dangerous. As a rough guide to the venom potential of a scorpion, look at the pedipalps, or claws. The less dangerous species such as *Pandinus imperator*, the Emperor scorpion, have large, strong pedipalps for grasping and crushing their prey. They depend on their strength to immobilise it. Members of the Family Buthidae, for example *Leiurus quinquestriatus*, the Israeli Gold, have narrow, delicate pedipalps. They depend on the venom in their sting to immobilise the prey as soon as it has been caught. Some such species, about 25, have the potential to kill a human. These, and in fact all scorpions, are greatly feared by the indigenous population of their country of origin.

Scorpions are much less demanding to keep in captivity than tarantulas. It has been suggested that they survive on neglect, although this is not to be recommended. They are secretive, nocturnal creatures, so are rarely seen during the hours of daylight. Unless a set-up is arranged using a red light where they may be persuaded to emerge during the day, they are not an ideal subject for display. If black light is used the large black scorpions will glow green.

As with tarantulas, these arachnids do not require a lot of space. They are rather lazy creatures and prefer to wait for their prey to come to them. A glass aquarium or plastic pet container with a secure, tightly fitting lid is required. A clear plastic food container 30cm. × 15cm. × 10cm. is adequate to house an individual of even the largest species. The substrate used is dependent on species to be housed and can be either Vermiculite, sand or forest bark. A stone or piece of bark should be provided as a retreat. Again, depending on species, temperature and humidity can range from tropical rain forest conditions to those found in desert regions or Mediterranean areas. One species, *Euscorpius flavicaudis*, has become naturalised in Britain. More detailed requirements are given under specific species. If heating is required by the species being kept, the same methods as those recommended for tarantulas can be used, bearing in mind that most scorpions will burrow. If a cable or pad is used it should only cover a portion of the container floor to allow the scorpion to choose an area in which it feels comfortable.

An essential item of equipment for the scorpion keeper is a pair of long tweezers or forceps, at least 25cm. long. It is not recommended that any scorpion species be handled, so tweezers should be used to pick them up by the segment just in front of the sting. The ends of the tweezers can be padded with foam rubber to minimise damage to the scorpion. This method is also recommended when collecting scorpions in the wild. Care must be exercised as scorpions can be very agile and can turn to use their claws to nip the handler. If this happens, make sure that the sting is not released. Another method of collecting or transferring scorpions is to use two clear plastic containers. One is placed over the sting and the other in front of the scorpion. If these are moved towards each other, the scorpion can neither sting nor nip. (Fig. 43)

The moulting process is critical for scorpions as a large percentage do not survive it. Sexually mature specimens do not moult, but the young must do so to enable them to grow. As with tarantulas, the relative humidity of the scorpion's environment is an important factor. It is best therefore to provide a variety of humidity levels within the container. Smaller species will mature after four or five moults, within a year, whereas larger species such as *P. imperator* will take seven moults over a longer period.

Scorpions can be overfed. Whereas a tarantula will refuse food if satisfied, a scorpion will overeat. The result is that the digestive system cannot cope with the surplus and it will die. They should only be fed once a week or fortnight when adult, and the young about twice a week. If the scorpion is unenthusiastic about prey introduced it is probably not hungry and should not be fed. Also if it leaves part of the item of prey, it has had enough for a couple of weeks. A sign of an overfed scorpion is if the membrane between the intersegmental plates and dorsal and ventral sides becomes distended. In a *P. imperator* the grey area along its sides is distended. This is usual in a gravid female, but unhealthy in a sub-adult or an adult which has not been mated. Scorpions can be fed on similar

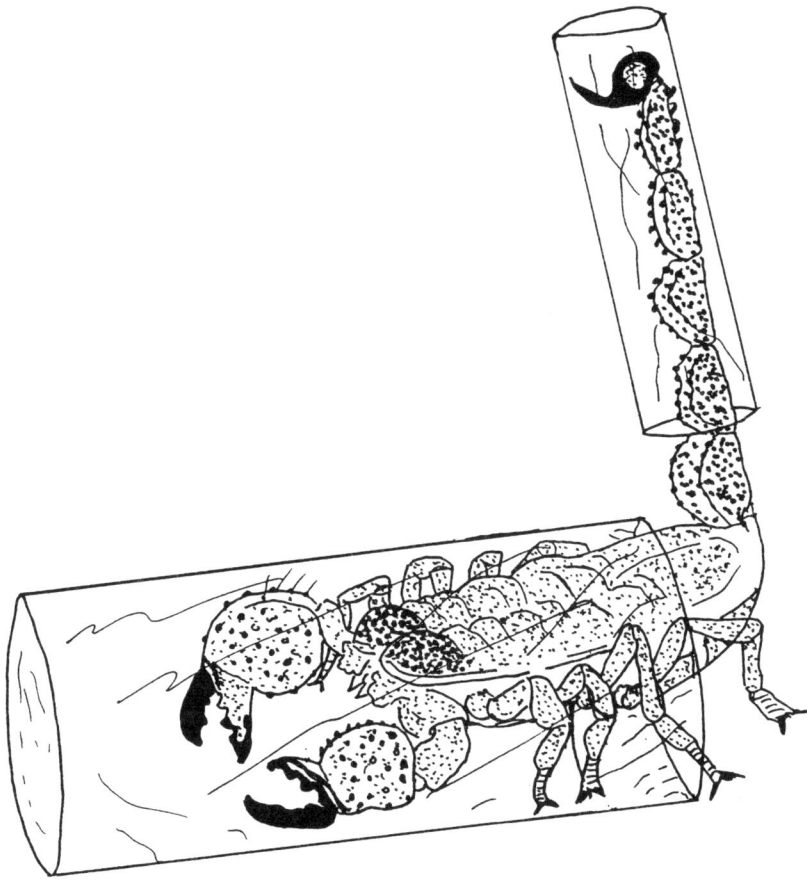

Fig. 43 A safe method of collecting scorpions.

prey to tarantulas, that is invertebrates such as locusts, crickets, flies, etc. They will also accept freshly dead prey, and a large locust may be shared by a number of communally living individuals such as *P. imperator*. Uneaten remains should always be removed to prevent infestation by mites. Water dishes are not essential as long as condensation occurs with the nightly drop in temperature. Spraying is beneficial to rain forest species.

Scorpions can be bred successfully in captivity. They have a unique way of mating in that the sperm is not transferred directly from the male to the female. The male grasps the female's pedipalps in his own and leads her around until he finds a suitable place to deposit his spermatophore. This is generally a smooth site such as a stone. He then manoeuvers her over the site until her genital opening is above the spermatophore. She then lowers herself onto it and transfers the sperm into her spermathecae. Evidence that mating has taken place can sometimes be seen where the scorpion has cleared an area of substrate on which to deposit his spermatophore. After a few months the female will give birth to live young which will then collect on her back. The young at this point are white. They will remain on the mother until they moult, after which they begin to disperse. The young should now be removed to avoid cannibalism. The rearing containers should be small enough to enable the young scorpions to feed easily. Emperor scorpions can be reared as a group with the mother as long as they are all adequately fed. The young can be seen to gather round a fairly large item of prey (a squashed adult cricket, for example) and feed communally.

***Pandinus imperator*** - **The Emperor or Imperial Scorpion**. This is the largest species of scorpion living today. It comes from the rain forest areas of West Africa where it lives in burrows on the forest floor. It is an impressive creature being glossy black with a faint greenish tinge. When kept under a black light they appear fluorescent, resembling alien creatures moving around in the dark. The size of adults can be from 15 to

Fig. 44 *Pandinus imperator.*

20 cm. though a variation in size does occur. An Emperor scorpion will take about seven moults to reach maturity, this requiring about four years. They can live for about eight years, but as captive breeding has not been in progress for very long the exact life-span is not known.

*P. imperator* is a fairly social scorpion and they can be kept in groups successfully provided enough food is given. It is wise to remove a female after she has given birth to prevent the adults feeding on the young – captive scorpions are not in their natural environment and cannot get away from each other. The young can, however, be safely left with their mother. It is difficult to introduce an outsider to the group as it will not be accepted by the majority. A substrate of peat and forest bark can be used allowing for the formation of burrows. Temperature should be around 26°C (78°F) and relative humidity about 80%. This scorpion is considered relatively harmless, but should not be handled as the sting can be painful. To sex them, compare the pectinal teeth underneath. Those on the male are larger than those on the female. (Fig. 44)

***Heterometrus spinifer* – The Asian Black or Forest Scorpion.** *H. spinifer* and other species of this genus are the Asian equivalent of *P. imperator*. They do not grow quite so large and are somewhat slimmer in build. They originate from the Far East and are found in Malaysia and Thailand. Their sting is considered to be more painful than that of their African counterpart and they should not be handled. Requirements are the same as for *P. imperator*.

***Scorpio maurus* – The Large Clawed Yellow.** This is a relatively harmless species of scorpion which can be found around the Mediterranean coasts, particularly in North Africa. There are a number of sub-species but they are generally orange/yellow in colour with heavy pedipalps. Adults are 6cm. to 7cm. in length. They are a burrowing species so this should be allowed for in substrate. In a collection site in Northern Morocco the burrows, as with those of all scorpions, were very distinctive in shape. They were oval to allow for the width of the pedipalps. The ground was hard clay and the burrows were about 20cm. deep. There were large communities with individuals living in close proximity to each other, both in woodland and on exposed slopes facing the sea. Humidity should be about 70% and the temperature range from 18°C to 23°C (65°F to 75°F). This is an interesting little scorpion to keep and it has been bred in captivity. (Fig. 45)

***Euscorpius flavicaudis*** This is the only species of scorpion to be found in Britain, where it has become naturalised. It probably arrived with imported goods from Mediterranean and European ports. It is rather

Fig. 45 *Scorpio maurus*. The Large Clawed Yellow Scorpion.

small for a scorpion, being only about 4cm. long. It lives behind bark and between bricks in walls. It can be kept easily in European homes because normal household temperatures are all that is required. It is not an aggressive species and not potentially dangerous, but it should not be handled.

*Opisthacanthus crysopus* - **The Yellow Legged Creeping Scorpion**. This is a small heavy bodied scorpion. It lives in burrows. It is a delightful little scorpion to keep as it is quite active. It requires temperatures around 25°C (78°F) and humidity of 75%.

*Liochelis australasiae* - **The Tiny Malaysian Brown**. This is a wood scorpion which lives behind bark. It is relatively easy to keep, although very small, being only 1.5cm. when adult. It has been bred in captivity. It needs a small container with bark or stones behind which to hide. (Fig. 46)

**DANGEROUS SPECIES** In the United Kingdom all scorpions of the Family Buthidae require a licence to be kept. Such species should not be readily available, and if offered for sale, the vendor should ensure the purchaser not only holds a licence, but knows exactly what he is buying. Some of these species may be seen and collected by enthusiasts on holiday, so every precaution should be taken to avoid being stung.

Fig. 46 Building block housing carved for Tiny Malaysian Wood Scorpions.

Fig. 47 Kenyan Orange Scorpion (Buthidae).

The characteristic of the Family Buthidae which have potent venom is the narrow pedipalps or claws. One of the most delicate, yet most deadly of scorpions is *Leiurus quinquestriatus*, the Israeli Gold. It is a pale yellow colour with a grey segment at the end of its tail. It is found around the Eastern Mediterranean where it lives under stones and houses. Its natural habitat is almost semi desert. The most common species in Europe in this Family is *Buthus occitanus*. This is found in southern Europe and north Africa where it lives under stones and rocks. It is a fairly large yellow scorpion growing to about 8cm. This is found in fairly dry areas.

Another potentially deadly species from north Africa is *Androctonus australis*. Along with other members of the genus these are known as Fat-tailed scorpions. *Centruroides species* from America are also dangerous.

All scorpions should be treated with the greatest respect. If in doubt about the venom of an individual, leave well alone. There are enough harmless species available which are in some respects more interesting to keep and observe.

Chapter Thirteen

# Other Arachnids

The Class Arachnida consists of 17 Orders which include Araneae (spiders) and Scorpiones (scorpions) as the best known and most commonly kept. There are other Orders which are less impressive in size, but equally interesting. They are rather more difficult to keep and few have been bred in captivity. (Fig. 48)

*Amblypygi* **- the Tail-less Whip Scorpions**. These are impressive members of the Class. They are found around the world, mostly within the tropics. In America they are most common south of Panama, they are widespread in South Africa and are also found in India, Borneo and New Guinea. They are often plentiful in the areas they inhabit and are found among rocks, in caves and frequently in houses. They have a flattened appearance with very long legs. The first pair are elongated and are not used for walking but are held in front as tactile organs. The pedipalps are also very long and these are used for securing their prey which they then move up to the chelicerae. They are also used for cleaning and preening the legs and body. Amblypygi can be kept in a plastic container or aquarium about 25cm. square with a close-fitting lid. The substrate can be sand, peat or Vermiculite and should be damp. These are rather secretive creatures so they should be given a piece of bark behind which they can hide. They can be fed on small invertebrates such as crickets or flies. Although rather gruesome, these animals present no danger to humans as they have no venom. They rely on speed to escape predators, running either forwards or backwards. Alternatively they may 'play dead'.

Amblypygi take about two years to reach sexual maturity and live for several years after that. The female lays about 20 eggs in an egg case which remains attached to the lower surface of her abdomen until the

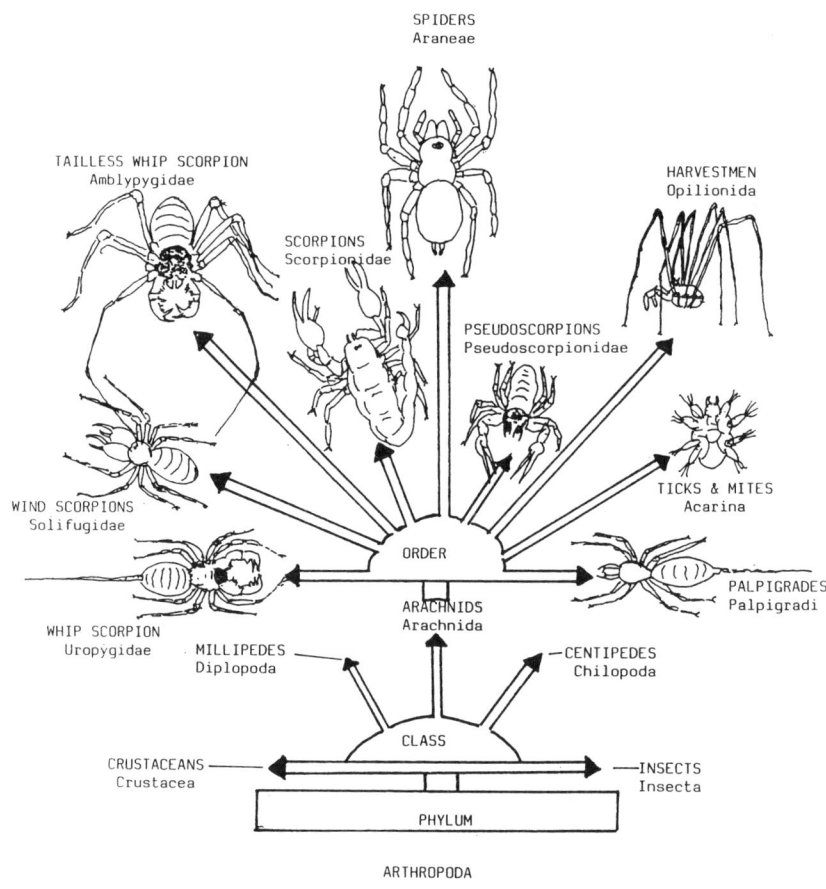

Fig. 48 A classification of Arachnids.

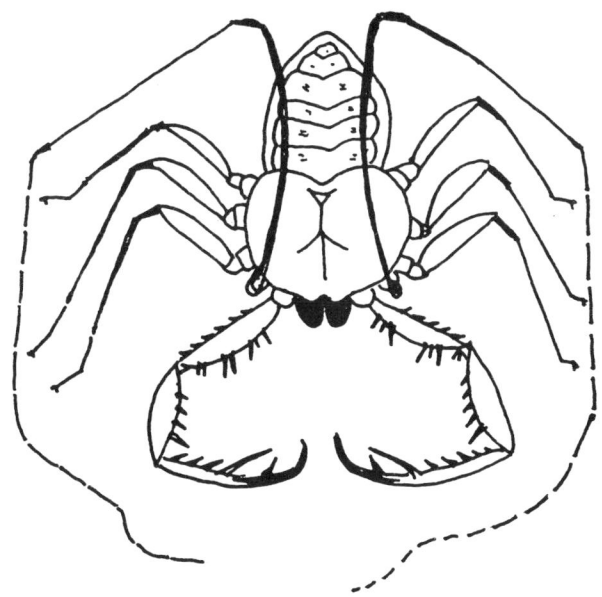

Fig. 49 Amblypygi.

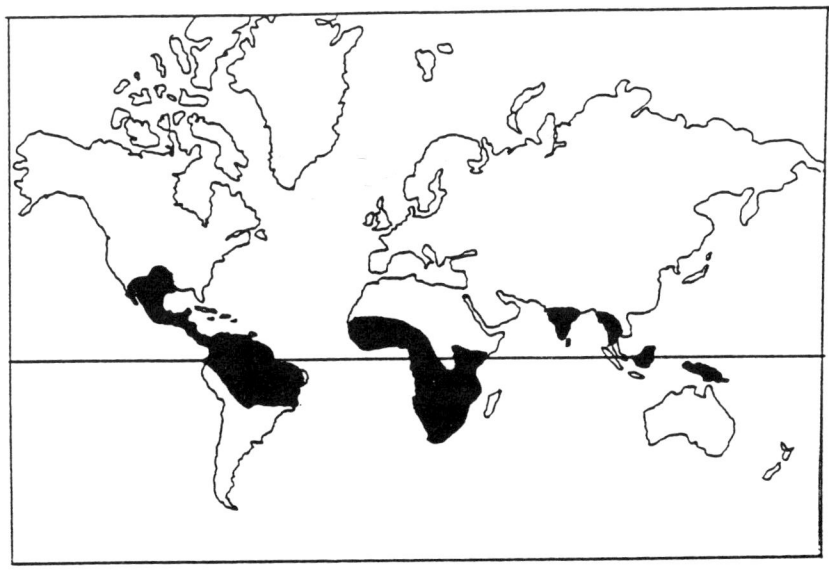

Fig. 50 Distribution of Amblypygi.

young hatch. They cling to her back for about two weeks until they moult. Then they should be separated. (Fig. 49 and 50)

***Uropygi*** **– The Whip Scorpions.** The Uropygi have features similar to the Amblypygi. The first pair of legs are used as tactile organs, although they are not quite so impressive. The pedipalps are also powerful appendages for grasping the prey and moving it to the chelicerae. Neither do they have venom. The most obvious difference is the tail. Although this bears similarities to the tail of true scorpions, it does not have a sting at the end. It is thought to contain chemical receptors. It is unwise to handle these creatures as they can excrete formic and acetic acids as a means of defense. Uropygi are found in the Americas, in the southern United States, in Central America and in the north-east of South America. They are also found in India and Eastern Asia. The only specimens recorded in Africa were found in The Gambia and these are thought to be descended from 'immigrants' travelling with imported goods. Uropygi can be kept in a plastic container, but this must be kept quite humid. They like to hide in a moist habitat, under stones or decaying logs. Some species are found in drier regions but these only emerge at night during the rainy season. They can be fed on small invertebrates.

Uropygi and Amblypygi do not copulate when mating but go through a complicated courtship procedure before the male deposits a spermatophore and the sperm is transferred to the female in a similar manner to the mating of true scorpions. (Fig. 51 and 52)

***Opiliones*** **– The Harvestmen.** Species belonging to this Order are found throughout the world. They are never very large but their legs appear disproportionately long. Their bodies have only one segment and they are ornamented with spikes. They have two eyes which are placed back to back. Of their eight legs, the second pair are indispensable. With these they can sense smell, taste and also perhaps sound. If they lose these legs they will die. In summer and autumn, Harvestmen can be seen clambering about in vegetation. They have proved difficult to keep in captivity. They need a diurnal variation in temperature. High humidity is required during moulting, but if this is maintained the new cuticle will not harden correctly and the animal will die. They therefore need a variation in humidity which is difficult to provide in a closed container. The lid should be made of fine netting to allow this variation to occur. If the substrate is kept damp they will be able to find the most suitable humidity levels within the container. The temperature at which they are kept should be appropriate to that of the country of origin. As well as hunting live prey, Harvestmen are also scavengers. As it is difficult to sex these creatures, a number can be kept together for breeding. Provided they have enough food and space, they should not attack each other.

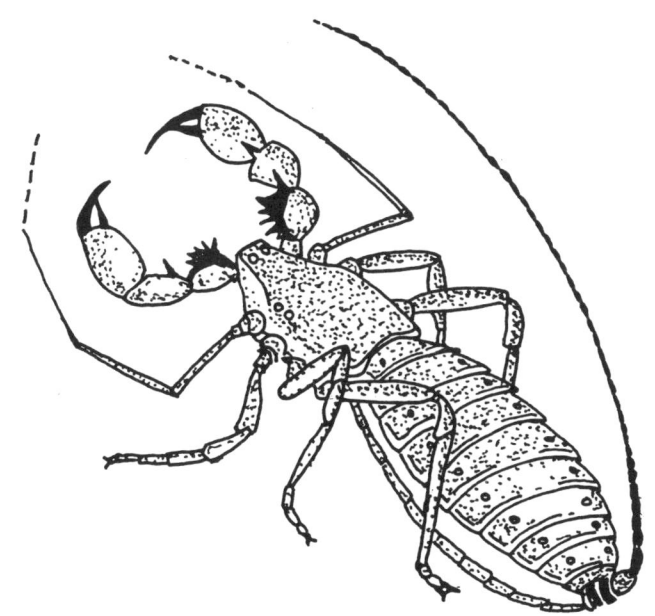

Fig. 51 Uropygi.

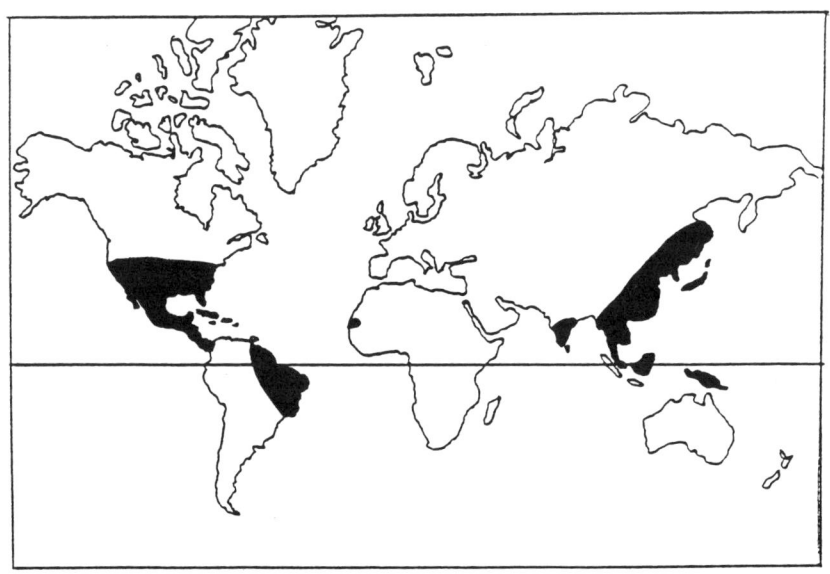

Fig. 52 Distribution of Uropygi.

When cannibalism does occur it is usually on dead or dieing animals. They are often found in quite large groups in the wild.

Within the Order Opiliones is the Sub-order Laniotores, which is the dominant group in the southern hemisphere. Members of the Family Gonyleptidae have been offered for sale in the United Kingdom, having been imported from Chile. They are interesting little creatures to keep and they breed well in captivity. In the wild they are found in dark, damp places such as caves, so the humidity should be quite high, about 70%. Care should be taken that they do not become infested with mites as these will eat the eggs which are deposited on the substrate. Gonyleptidae have an arrangement of spikes on their bodies and fourth pair of legs, probably as defense. They also emit quite a strong odour to deter predators. A number can be kept together and they readily produce eggs. (Fig. 53)

***Acari*** **– The Mites**. Much has been said so far about the undesirability of mites. The parasitic species can harm tarantulas and other arachnids, but some can cohabit without being a danger. It is usually only when the spider dies that there appears to be an explosion in numbers, or if the temperature and humidity are high and food is left to decay. A heavy infestation of even non parasitic mites can cause distress to a spider and if they are enclosed in an egg-sac they will destroy the eggs. All mites should be avoided. They are generally so small that they are of no interest to the animal keeper in their own right, but some species are of economic importance. One such species is *Metatetranychus ulmi* which attacks fruit trees and other crops, sucking out the plant's juices. They have become a major problem since pesticides have been used to kill other insects, the predators of the mites.

The only member of the Order Acari of harmless interest to the arachnologist as a pet keeper, is the Giant Red Velvet Mite (Dinothrombium) which are about 1cm. long. They are the largest of the Acari. Their bodies have the exact texture of velvet. They are mainly found in desert and semi-desert areas and also in some more humid parts of the tropics. After heavy rains they emerge in huge numbers, carpeting the ground in velvet. They are not easy to breed in captivity because the larval stage is parasitic and requires a locust to which it can attach itself. The adults will feed on almost anything. They are very catholic in their tastes and will eat living or dead plants or animals. Mites as arachnids are interesting but due to their size and reproductive cycle, are not very practical to keep. (Fig. 54)

***Pseudoscorpiones*** **– The False Scorpions**. These are very small, most being less than $\frac{1}{2}$cm. in length. They resemble true scorpions apart from

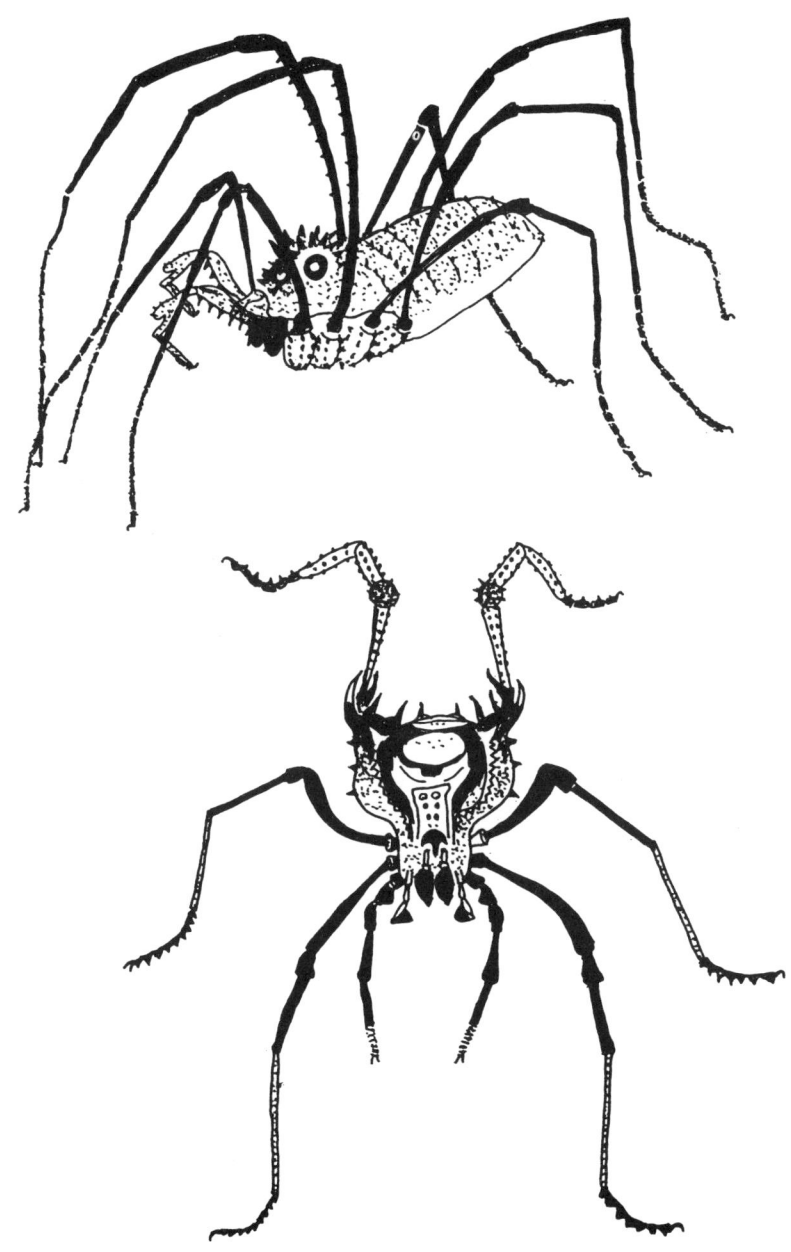

Fig. 53 A Harvestman and a Chilean Laniotore.

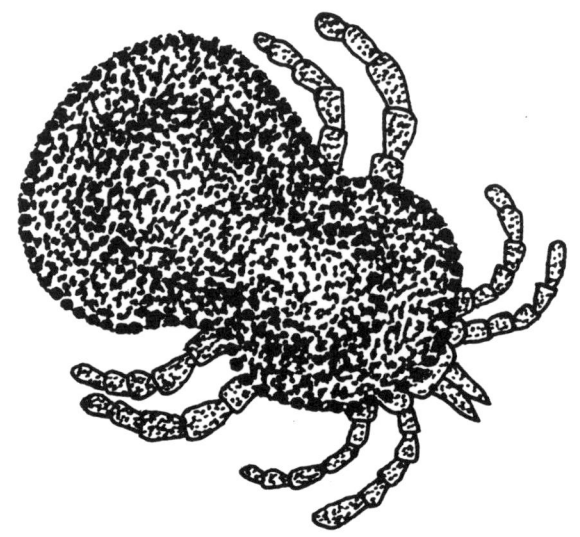

Fig. 54 A Red Velvet Mite.

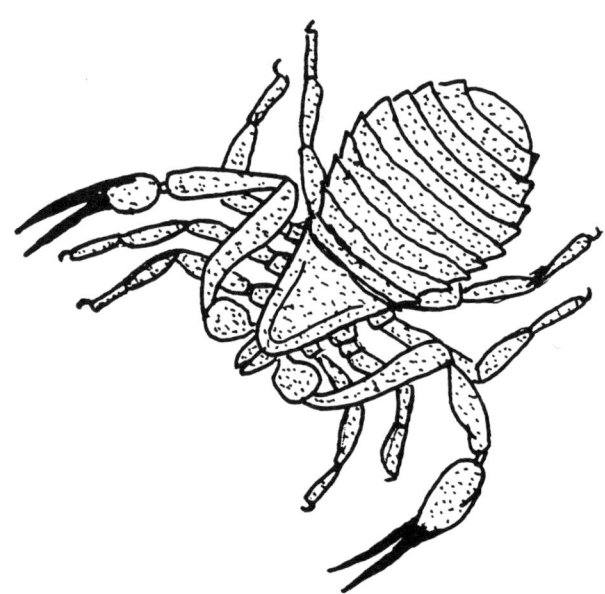

Fig. 55 A pseudoscorpion.

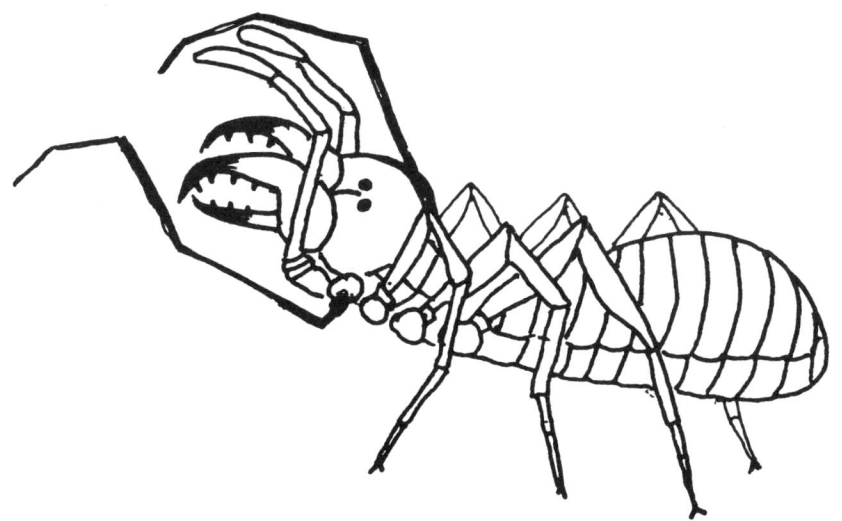

Fig. 56 A solifugid.

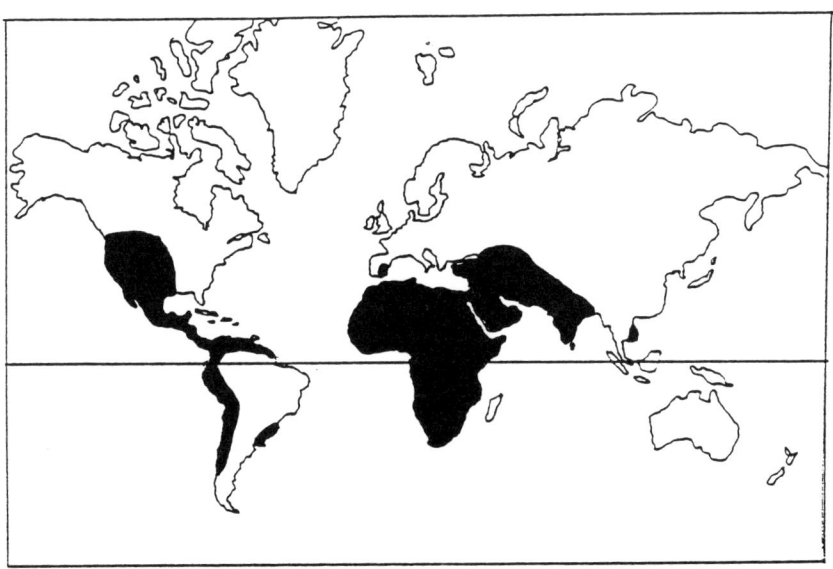

Fig. 57 Distribution of Solifugids.

the lack of a tail and sting. In Britain they live in leaf litter and manure heaps where they find tiny invertebrates on which they feed. They are found throughout the world except for the frozen wastes of the Arctic and Antarctic. They have many interesting features, but unfortunately their size makes them more suitable for the specialist. (Fig. 55)

*The Solifugae* – **Camel spiders, Wind spiders, Sun spiders**. The solifuguds have a number of common names depending on their country of origin. They are much sought after by the hobbyist but are rarely available. They are confined mainly to tropical and sub-tropical regions but one species occurs in southern Spain. They are found over almost the whole of Africa, stretching across the Arabian peninsula to India. In America they cover the western United States from the Canadian border through Central America and southwards as far as Terra del Fuego. It is therefore a misconception that they are desert arachnids. Some species do originate from desert areas where they are nocturnal, but many diurnal species live in forests and grasslands. In many countries Solifugids are greatly feared by the native inhabitants, but this fear is unfounded. It may be their speed, their size, their resemblance to spiders or their massive chelicerae or jaws which cause this fear. Although they possess no venom, they look threatening. The body surface is covered in sensory hairs up to 50mm. in length. Organs of taste, smell and hearing are present on the feet and mouthparts. There are also other sensory organs called malleoli at the base of the last pair of legs, the exact function of which is not known. The feet have a series of stiff hairs and spines, allowing solifugids to run rapidly over loose sand. The diurnal species have dark markings on the body and legs and tend to be smaller than the nocturnal species, which are yellow to brown in colour and may reach 75mm. body length. Mature males can be identified by the species specific shape of horn-like extensions on the upper jaws. These structures are absent in female and immature specimens.

Solifugids have proved difficult to keep in captivity when they have been imported, and they have not been succesfully bred. Most that have come into the United Kingdom have originated from the desert areas of Egypt and the Middle East and although attempts have been made to reproduce this environment, they have soon died. They are solitary creatures living in shallow retreats scraped out under rocks and logs. If a male and female are mated she will produce about 20 eggs and guard the egg-sac until the young hatch. They should be separated when they moult. It is unfortunate that there has been so little success in keeping and breeding Solifugids. Obviously much more information is needed about them. (Fig. 56 and 57)

Chapter Fourteen

# Collecting and Conservation

Collecting spiders and other arachnids can provide a lot of useful information on their natural habitat and lifestyle. Most of the information available on the environment in which a species lives is very general, usually only giving the country of origin and whether the collection site was rain forest, scrubland, etc. Details of the micro habitat are important, such as temperature, humidity and depth of burrow. Other interesting information can be gathered such as the preferred food of the spider by noting the remains, or the proximity to others of the same or different species. An expedition purely for collecting spiders can be expensive and is not practical for most people. More feasible is to combine the interest with a family holiday. Whatever the case, careful preparation is needed to ensure success.

One of the major considerations should be the law. The Convention on International Trade in Endangered Species of Wild Flora and Fauna (CITES) has restricted the movement of many species across international borders. These regulations apply to private collectors as well as traders and have been drawn up to protect endangered, threatened and vulnerable species. There is, however, only one spider included, in Appendix II, threatened species. It is the Mexican Red Knee (*Euathlus smithi*). Other invertebrates are included, so the list should be checked against what has been collected.

As far as customs regulations are concerned, there are no restrictions on importing live spiders into the United Kingdom as they do not carry any disease which can be transmitted to humans or other animals and do not pose a threat to crops. For the last reason certain beetles are prohibited. It would be very unwise to try to evade the customs regulations as the penalties for smuggling are severe and everything obtained outside the country, legal or otherwise, will be confiscated. To avoid problems it is best to pass through the Red Channel and declare any livestock. Have the specimens labelled and a separate list of all items as well as a copy of the CITES list. Looking professional can help to avoid lengthy investigation.

It is advisable to be aware of the regulations within the country to be visited. Certain countries have no restrictions on collecting invertebrates, others have strict rules prohibiting their export whereas some will give permission for a limited number to be collected. Information can be obtained from Embassies, High Commissions and Tourist Offices.

Once a destination has been decided, research with regard to temperature, rainfall and natural vegetation should be carried out. Particularly in tropical regions, the date of the rainy season is important. If there is a distinct dry period, very little will be found at this time. It could be compared to the temperate winter when, due to adverse conditions, the invertebrate population is reduced and awaits the coming 'spring' – the rains. The best time to collect is after the rains have started, or directly after them. Many spiders spend the dry season deep in their burrows to avoid dehydration and because there is a shortage of invertebrate food for them. Others, the annual spiders, pass this period within the egg-sac which will hatch when the humidity rises. Some knowledge of the type of spiders that could be found will be useful. The natural vegetation of the area will give some indication. Orb weavers will require something stable to which they can attach their web and arboreal species of tarantula require a fairly densely wooded area. More open ground is suitable for some burrowing species, although some cover in the form of rocks or bushes is important.

It is necessary to think about containers and how they should be packed. Airlines do not approve of carrying livestock as hand luggage, so they must go into the baggage compartment. Careful packing is essential as this part of the aircraft becomes very cold at high altitude. A polystyrene box is ideal and can be packed inside a suitcase with clothing as extra insulation. For travelling, the smallest containers possible should be used as this prevents the spiders from being damaged when the luggage is loaded and unloaded. After collection the spiders need quite a large area so containers which fit inside each other are useful as they do not take up much space when not in use.

Travelling by car is not so restricting as far as space is concerned, but extra care must be taken to avoid extremes of temperature. The interior of a car can become very hot and heat will kill spiders faster than cold. Again a polystyrene box will provide insulation, but should not be left in the sun.

To collect spiders, other equipment is needed. Burrowing spiders, as well as scorpions, have to be dug out. This must be done very carefully to avoid harming the animal. A length of flexible plastic tubing (about 5mm. in diameter) to insert into the burrow to establish its direction is needed as well as a small trowel with which to dig. A sweep net can be used for collecting specimens from vegetation as well as a white plastic sheet onto which the net can be emptied. For smaller specimens a pooter is needed. If scorpions are to be collected it is essential to carry tweezers. A supply of labels and a field note book should also be taken. To collect data on the habitat, take a thermometer and humidity gauge and a tape to measure depth of burrows, etc. If possible obtain a large scale map. As well as providing detailed information on the area an accurate map reference can be worked out pinpointing the exact location that a specimen was collected.

To select a suitable collecting site certain factors must be considered. The site should be undisturbed, particularly when looking for relatively long-lived tarantulas. Not many virgin areas remain in the world, so an area that is unsuitable for cultivation should hold a community of spiders. Ideally there should be rocks or other cover such as fallen trees. The rocks should be of a size which can be turned over fairly easily. Those which are deeply embedded in the soil or very heavy will not be used by invertebrates as they will not be able to get under them. Small rocks which are easily moved will also be avoided by these animals as they will not provide secure cover for them. Small stones, such as scree, which move frequently will also be unproductive. Particularly in areas where there is heavy rainfall there should be a slight gradient to provide run-off. Assuming the collecting site is to be a rocky hillside, try to find a shady 'base camp' where specimens can be safely left. Many rocks may have to be turned over before anything is found. However, once one specimen has been discovered, spotting them becomes easier, for example the size and position of favoured rocks and tell-tale signs such as web protruding. It is very important to replace even unproductive rocks. The area should be covered systematically, a small patch at a time, and specimens put in the shade as soon as possible. Close watch should be kept on the 'base camp' in case it becomes exposed as the shadows move.

Rocks are not essential for a good collecting site. Some burrowing species of tarantula will inhabit banks and the sides of hillocks. In this case it will be the entrance to the burrow which will be seen. Often there are traces of the web which is used to line the burrow. During the day spiders will often web across the entrance, opening the burrow when they wish to feed. This web is a sign that the burrow is occupied and it is a deterrent to some predators. Extracting tarantulas from their burrows must be done very carefully to avoid damaging or even killing the creature. It is not practical to wait for the spider to come out as this generally happens at night and the slightest movement will frighten it. Many species do not leave the burrow completely, but wait at the entrance for prey to come close enough to be pounced on. Digging is usually the only method that can be used. To avoid losing the path of the burrow, the flexible plastic tubing should be inserted carefully. Using a small trowel the earth around the burrow can be removed, so that what is left is a hole with the burrow intact in the centre. With the tube held in place the earth immediately surrounding the burrow can be removed. With luck the spider will be in a cell at the base. It will either be very frightened or very angry, so care must be taken when it is being put into a suitable container. Scorpions can also be collected in the same way. Their burrows have a characteristic oval opening to accommodate the pedipalps and sometimes have a pile of excavated earth around them. A lot of patience is needed when collecting in this way, whether it is scorpions or tarantulas, as the animals can so easily be damaged by careless digging.

It is possible to encourage a spider or scorpion to come out of its burrow without digging. During the dry season scorpions in particular can be extracted by wetting the end of a piece of palm frond or grass with saliva and inserting this into the burrow. When the scorpion grasps the end to drink, it can be carefully pulled out. Tarantulas, especially the more aggressive species will emerge if irritated in a similar way. Many species construct an escape route, a back door, whether they construct a burrow or merely a scrape under a rock. Before the spider is disturbed all possible exits should be noted. It may only be necessary to give an impression of a predator, with a stalk of grass, at the 'front door' for the spider to try to make its escape through the 'back door'.

In areas of dense undergrowth arboreal species can be found. Although these species generally construct a tube web in which to live, these are well camouflaged. They can be behind loose bark or in the folds of leaves. Banana and pineapple leaves are a favourite of *Avicularia avicularia*, the Pink Toed tarantula. It is relatively easy to extract these spiders by pulling the leaves apart and exposing the web.

Although the majority of tarantulas are sedentary creatures and do not readily leave their burrow or web, the mature male is an exception. The aim of a mature male, in his short life, is to locate and mate with as many females as possible. It can be assumed, therefore, if a large tarantula is wandering about, generally as dusk approaches, it is a male in search of a female. Unless a female is available for mating there is little point in collecting these males except for observation.

Dipluridae species, with their characteristically long spinnerets can be found in leaf litter or amongst the branches of shrubs. These webs are usually well camouflaged with debris such as dead leaves. These spiders can be difficult to catch due to the size and complexity of their webs. At any sign of disturbance they can choose one of many escape routes and amongst dense vegetation can easily make their getaway. A sweep net is invaluable in this case. The entire mass of the web including the debris is scooped up into the net. Whilst still in the net, unwanted material can be sorted and discarded, leaving, hopefully, the spiders. These will try to run up the sides of the net but can be easily shaken back until they can be put into a container. Due to the size of the webs they often contain a community of spiders, sometimes of more than one species. As well as dipluridae, araneomorphs can be collected at the same time. Valuable information on interdependence can be obtained, and on communal living, from these webs.

Trapdoor spiders present the problem of the excellent camouflage of their burrows. These spiders may never, during their lifetime, leave the original burrow. They enlarge it as they grow. The trap door is a perfect fit and is extremely difficult to spot. Whilst the spider is in its burrow, this door is firmly closed and flush with the ground. It is constructed of web and earth and is camouflaged with sticks, moss and leaves. It is possible, once a burrow has been located, to dig the whole thing out intact, with the spider inside. This is a kinder way to collect these species as they will be undisturbed and the burrow can be accommodated at home by burying it in soil. The spider can then be observed.

The indigenous population of an area being visited can provide useful information on the location of tarantulas. Children can often be persuaded to collect spiders in return for a few coins. They find it difficult to understand why anyone should want to collect spiders, especially as they are considered by many of them to be dangerous, but will humour the collector.

Many interesting araneomorphs, or true spiders, can be found, particularly in tropical regions. The giant Golden Silk spider, Nephila, is easy to spot in the right habitat, hanging from their webs. As stated earlier, collecting adults is not very successful, but egg-sacs can be easily transported. Giant Huntsman spiders, *Heteropoda venatoria*, are often found in dwellings and can be 'bought' for a few coins. Jumping spiders, Salticidae, and other orb weavers are also interesting to collect.

Once the specimens have been collected, they must be housed and fed until they can be brought home. A great problem is ants. Many species will prey on the captive spiders and they seem to be able to get through impossibly small spaces. To try to overcome the problem put the containers in the polystyrene box. Set this box on four small blocks or corks on which has been put some ant killer. The ants will not be able to climb into the box. Alternatively float the box in a basin or bath of water so that it is not touching the sides. Be careful that the ant killer does not come into contact with the spiders and spray should not be used. Ensure that the spiders are kept at a reasonable temperature.

The tarantula keeper, when collecting, will be more interested in observing than in pure scientific facts which can only be obtained from dead specimens. However, if a specimen dies after capture, it need not be wasted. It can be preserved for future use, either to set it or examine it. To prevent decomposition, alcohol or formaldehyde should be used, but these may not be practical for the amateur as they require a licence for purchase. A more practical alternative is surgical spirit although this does evaporate quickly so needs an airtight jar. If nothing else is available gin, as high a proof as possible, can be used in the short term.

Conservation is a very important factor which must be born in mind whilst collecting spiders. The populations of many species have been depleted in their natural habitats through indiscriminate collecting. The effects of this can be seen in many of the specimens being imported into the United Kingdom. Whereas large mature adults were being imported, many of the tarantulas available now are either immature or runts. In some sites all the mature adults have been removed and even if left alone now it will take many years for these colonies to re-establish themselves successfully. Many countries have now forbidden the commercial export of their native fauna.

The amateur collector, doing so for his or her own interest, is unlikely to cause such large-scale damage. However care should be taken not to take too many spiders from any one site. If the object is to breed the spiders in captivity a maximum of ten mature females would be more than adequate. A gravid female or an egg-sac will provide future males while any immature specimens could well mature into males. These numbers will ensure a large enough gene pool to keep the species going in captivity.

One of the greatest problems that conservationists face is the destruction of natural habitats. The destruction of rain forests is reducing the area available for many species, some of which may even now have been wiped out before they have been recorded and so are lost for ever. Urbanisation and intensive farming have also reduced natural habitats and pesticides and pollution have depleted other species. Some species, however are opportunists and have adapted to 'village life'. The scorpion, *Leiurus quinquestriatus*, lives successfully in Bedouin villages in Egypt and the tarantula, *Pterinochilus murinus*, uses the ridges in corrugated iron roofs or the space under floors of shanty town huts in Africa. These, like the urban fox, have adapted successfully, but many species are unable to do so. The more that can be learnt about their habits, the better the chance will be of preserving them in the wild, as long as some of the habitat remains. Awareness of the problem is the first step to solving it and the amateur arachnologist has a part to play by captive breeding, observations and, most important, not over-collecting.

# SPECIES FILE

# North and Central American Species

*Euathlus albopilosa* (Valerio 1980) – The Curly Hair or Woolly Tarantula. A freshly moulted, well cared for *E. albopilosa* is a very attractive tarantula. It is a uniform brown on legs, carapace and abdomen and is covered with light tan/orange hairs. It is the hairs which give it its common name because they have a distinct curl. It has been described as a tarantula with a perm! It is a fairly docile species, but it is capable of moving quite fast, so, for the sake of the spider, it is not recommended for handling. However it is a good species for the beginner as it is quite hardy. Captive bred specimens are available in the United Kingdom and spiderlings are relatively easy to rear and grow quite fast. The females reach maturity in about three or four years and the males in slightly less. A well cared for female can produce an egg-sac containing 700 young, a fact that must be considered before mating – can such a large number of spiderlings be cared for adequately? *E. albopilosa* is found in Honduras and Costa Rica in areas of tropical rain forest, therefore it requires fairly high humidity (75%–80%). It is a burrowing spider so a deep substrate should be provided.

*Euathlus smithi* (Cambridge 1897) – The Mexican Red Knee. This is probably the best known of all the tarantulas, and one of the most striking in appearance. For many years they were exported for the pet trade with the result that there is now concern for their survival in the wild. They are at present on CITES Appendix II which means they cannot be moved from one country to another without accompanying paperwork. The species is now being bred successfully in captivity. One reason for *E. smithi*'s popularity is its hardiness and docile nature. It will tolerate, as an adult, less than perfect conditions and in the past has survived a 'hot desert' environment in captivity. *E. smithi* is a very attractive species. It has a black abdomen which is covered with bright red hairs. The legs are brown with bright red/orange 'knees' which are bordered with cream. The carapace is tan/cream. A stressed or frightened individual will readily throw hairs, resulting in a bald patch on the abdomen. These hairs can be quite irritating if the owner is exposed to them over a lengthy period. *E. smithi* originates from the desert regions of Mexico where it lives in deep burrows in banks. During the dry season it will spend most of its time in the burrow where it is cool and damp so, although it is a desert species, it should NEVER be allowed to dry out. Although it is difficult to get hold of this species it is ideal for the beginner. Due to the efforts of members of the British Tarantula Society and the Zoological Society of London, a number of egg-sacs have been produced in recent years. The resulting spiderlings grow slowly and will take 5 to 6 years to reach maturity, but an adult female will live for over 20 years.

A very similar species to *E. smithi* has been imported recently. It also has red 'knees', but is a distinct species due to anotomical differences. It has been sold under the name Flame Knee (no scientific name) and as *E. smithi*. Little is known at present about this species except that it is only found in one small area of Mexico. There has been a lot of controversy over its true identity.

*Euathlus vagans* (Ausserer 1875) – The Red Rump or Black Velvet. *E. vagans*, when adult is a large attractive spider. It is basically black with the abdomen covered in reddish brown hairs. The legs and palps are velvety black. A freshly moulted specimen can be quite striking in appearance. There is some colour variation depending on the site of origin, some individuals having less intense black hairs on their legs, but this makes them no less attractive. There can also be some variation in the colour of the carapace, ranging from black to mid-brown in some individuals. As with the majority of New World species, *E. vagans* has urticating hairs on the abdomen and readily throws these if frightened, resulting in a bald patch. However, if treated with care and not stressed this species settles well in captivity and will retain all its hairs between moults if treated correctly. This species has been bred successfully in captivity and individuals of various sizes are available. The spiderlings are relatively easy to rear and at all stages of development *E. vagans* can be recommended for the beginner. It is a fairly docile species, rarely showing aggression, but quite active compared to other species recommended for the beginner. In the wild they are found in Mexico and Guatemala where they live in deep burrows in scrubland. Average temperature and humidity should be maintained. There are other species very similar to *E. vagans* which are sold under the same common name. If breeding is intended, positive identification of species is essential.

*Euathlus emilia* (White 1856) – The True Red Leg or Painted Tarantula.

This species has very distinctive markings. The abdomen is black with quite long orange/red hairs. The carapace is tan with a triangle of black, the apex of which is at the foveal groove and the base at the chelicerae. The legs are particularly attractive being black except for the tibia on each which is red. *E. emilia*, although an attractive spider is not considered suitable for the beginner. There is some question as to the toxicity of its venom. It is thought to be more potent than that of other species, but there is no evidence to substantiate this. It is not a species which can be readily handled as it is quite nervous. It readily throws hairs if aggrivated. *E. emilia* has been bred in captivity but it is rather slow growing. In the wild this species is found in semi desert scrub areas in Mexico and Panama. It requires slightly lower than average humidity (65%) and average temperature, with a substrate in which it can burrow.

*Euathlus mesomelas* (Cambridge 1897) – The Costa Rican Red Leg. There is quite a variation in the colours within this species. Basically the spider is black with reddish orange markings on its legs. It is these markings which vary from barely noticeable to quite vivid and covering much of the leg. Although the abdomen is covered with short velvety hairs it still uses these as a defense and throws them. The result is a pink bald patch if it is stressed. As the time for the next moult approaches the abdomen becomes lighter and the hairs turn brown. A dull brown *E. mesomelas* is therefore nearing a moult, particularly if the bald patch is turning black. This species can be quite aggressive and has long legs so can move fast. It is fairly hardy. It has not yet been bred in captivity to any great extent, but specimens are sometimes available. A female takes around five years to reach maturity. The natural habitat of this species is the rain forests of Costa Rica.

*Aphonopelma seemanni* (Cambridge 1897) – The Costa Rican Zebra. *A. seemanni* was, at one time, a readily available species, but recently few have been exported. Attempts are being made to breed it in captivity. Unfortunately much more information is required on their natural habitat and life patterns as they appear reluctant to mate. Matings which have so far occurred have failed to produce a fertile egg-sac although gravid imported females have eventually produced spiderlings. The adult female can be a very striking spider. The abdomen of a freshly moulted specimen is dark brown and is covered with orange/tan hairs. The carapace is brown. The underside is pinkish fawn. The legs are its most attractive feature, being dark brown with cream longitudinal stripes. Many of the imported specimens were mid-brown in colour and generally appeared dull. The change on moulting is considerable and if the relative humidity is kept quite high, the darker colour remains until the next moult. The male of the species is completely black, losing all traces of the stripes at the final moult. These spiders are found over a wide area of Central America, in Costa Rica, El Salvador, Guatemala, Nicaragua and Mexico, and regional variations do occur with some Zebras having a blueish tinge on moulting. Temperature and humidity should be higher than average, 27°C and 80%. Zebras will often go for quite long periods without eating before a moult - seven months has been known. They can be rather skittish so handling should be avoided.

*Aphonopelma chalcodes* (Chamberlin 1939) - The Palomino or Blonde. This is a much sought after species, but unfortunately is difficult to obtain. It is light brown and the abdomen and legs are covered with golden hairs. A freshly moulted specimen is somewhat darker but soon lightens to live up to its common names. A mature male is dark brown after his final moult. As a species it is fairly docile and generally quite inactive. It can go for long periods without eating. Although apparently docile, it can move surprisingly fast if frightened, so it should not be handled. Attempts are being made to mate the Blond with wild caught males, so captive bred spiderlings should be available. It is a slow growing spider which may take 5 to 6 years to reach maturity. This indicates a long life-span. The natural habitat is scrubland in Mexico and Arizona and it is a burrowing species. Average temperature and humidity are adequate, with a deep substrate. *A. chalcodes* tends to blend in with Vermiculite due to its colour, but it looks particularly attractive on a darker substrate.

*Psalmopoeus cambridgei* (Pocock 1895) - The Trinidad Chevron. *P. cambridgei* is an arboreal species, making its retreat in folded leaves and behind bark. It makes a tube web and compared to ground dwelling species it is very flat. The female is olive green with black chevron markings on the abdomen. Adult males have less striking patterns but longer hairs on their legs. Also they are of a lighter build and can move extremely fast. This species is fast growing, with the males reaching maturity in a year. This indicates a shorter life-span with the females living for 6 to 8 years. These tarantulas are definitely not to be handled. They are fast and aggressive, but interesting to keep as they are more active than ground dwelling species. They are masters of escape, squeezing themselves through unbelievably small spaces, so their container should have a secure, tight fitting lid. The natural habitat of *P. cambridgei* is the tropical rain forest of Trinidad, so temperature and humidity should be on the high side. Provision should be made for them to build their silken tube by placing cork bark or twigs in the container. Although aggressive to humans, these spiders are reasonably tolerant of the opposite sex and readily mate. Many captive bred spiders have been produced in Britain.

*Hapalopus incei* (Cambridge 1898) - The Olive Brown Trinidad Spider. This is a small species of rather nondescript appearance, but is interesting to keep. It constructs a retreat out of web and is fairly active. It is a brown spider with similar chevron markings to those of *P. cambridgei*. Due to its small size it has only been brought into the United Kingdom by amateur arachnologists, therefore is rather rare, but in its native Trinidad it is very common. It is fast growing so short lived, but is much more active than most of the larger species. In Trinidad the natural habitat is tropical scrub so it does not require high humidity. It should be provided with bark to which it can attach its web.

*Phormictopus cancerides* (Latreille 1806) - The Haitian Brown. *P. cancerides* is a very handsome tarantula, being one of the larger species. It is dark brown with some golden hairs on its legs. It has a large leg span. When mature the adults have an iredescent pinkish sheen on the carapace which is particularly noticeable on the mature male. This is definately not a species to be handled as it can move extremely fast and tends to be aggressive. *P. cancerides* has been bred in captivity and the spiderlings are quite fast growing. They feed well and do not have long periods of fasting as other species do. Mating these spiders can be traumatic as they seem to be very violent towards each other. The natural habitat of this species is rain forest with high humidity (80%) and they are widely distributed throughout Haiti, Puerto Rico and the Dominican Republic. This is not a beginner's spider due to its aggressive nature, but is interesting for those with some experience.

*Phormictopus nesiotes* (Chamberlin 1917) - The Golden Cuban Brown. This species is similar to *P. cancerides* but is smaller. It is dark chestnut brown with golden brown and grey hairs intermixed, giving it a greyish hue. It is also found in the northern Caribbean, in Cuba, so the two species could be closely related. There are few of this species in Britain at present but it is hoped to breed them in the future. The natural habitat is rain forest so quite high humidity is required.

*Cyclosternum fasciatus* (Cambridge and Valerio) - The Tiger Rump Doppel-ganger. This is a beautifully marked little tarantula. The legs are black and the abdomen is striped in black and pinkish orange. The carapace is an iridescent pinkish colour. This species is very similar to *Metriopelma zebratus*, the most obvious difference being in mature males where *M. zebratus* has no tibial spurs (mating hooks). Otherwise it is difficult to distinguish between the species. *C. fasciatus* originates from the rain forests of Costa Rica so requires 80% humidity. They have been bred successfully in captivity and are easy to rear provided food of the correct size is available. The spiderlings are very small so even micro crickets are too big for them to manage. The spiderlings soon begin to show their adult colours and take about three years to reach maturity. Due to their speed these spiders should not be handled.

*Metriopelma zebratus* (Banks 1909) - The Costa Rican Sun Tiger Abdomen. *M. zebratus* is very similar to *Cyclosternum fasciatus* except for the absence of tibial spurs in the mature male. It is a very attractive tarantula living up to its common name with orange/pink and black striped abdomen. The legs are black and the carapace is of an iridescent pink colour. The natural habitat is the rain forests of Costa Rica so fairly high humidity is required. This species has also been bred successfully in captivity but the spiderlings are very small and need appropriately sized food. If breeding is proposed, care must be taken to ensure that the male and female are of the same species due to the similarity with *C. fasciatus*. They are rather aggressive in mating with the female often attacking the male. It is a fast moving species so handling is inadvisable.

*Dugesiella epicureana* (Chamberlin 1925) - The Yucatan Rust Rump. This is a small but sturdy species from the tropical rain forest area of the Yucatan peninsula in Mexico. Although a uniform brown colour, it is an attractive little spider due to its shape. A well fed adult has an abdomen which is almost round. They are quite docile and can be handled. These have been bred in captivity but are a slow growing species, which indicates a relatively long life. The spiderlings on emerging are quite small so this should be born in mind if breeding is proposed.

*Citheracanthus crinirufus* (Valerio 1980) - The Costa Rican Blue Front. This is one of the smaller central American species which has been available in Britain. It is a sturdy little spider but can be quite aggressive. It is found in the Pacific coast tropical mixed forests of Costa Rica, so requires 80% humidity. The carapace is dusky grey with paler, reddish fringe hairs. The abdomen is dark grey with short reddish grey hairs. The legs are brownish grey. The distinguishing feature of this tarantula is that the chelicerae have a metallic blue sheen in natural light. This is particularly noticable in a freshly moulted specimen. Unfortunately the only 'captive breeding' was an egg-sac from a gravid wild caught female. The resulting spiderlings were easy to rear but the males matured before the females, so it would appear that captive breeding will be impossible unless new wild stock is made available. This is not a species that can be handled, and although easy to keep, is rather aggressive for the beginner.

# South American Species

*Avicularia avicularia* (Linne 1758) – The Pink Toed Tarantula. *A. avicularia* was the original 'bird eating' tarantula, its name being derived from the Latin for bird – avis. It is reputed to have been seen in the early 18th century eating a small humming bird. It is, however, one of the more docile species as well as one of the most commonly available. It is found in the rain forest areas of South America from Brazil to Guyana, Venezuela and Trinidad and has adapted itself to life with humans by utilising the corrugated iron and straw roofs of houses. The indigenous population would be well advised to encourage these spiders to share their homes as, apart from not being aggressive, they consume a large number of insects. *A. avicularia* is an arboreal spider. In its natural habitat it constructs white tubes of silk in leaves such as banana and pineapple. It is capable of taking small birds, but its chief prey is more likely to be invertebrates and perhaps tree frogs. Captive specimens often prefer flying food, but they will leave the web to catch crickets. Because it is arboreal in nature it will not readily drink from the substrate or a water dish. The substrate should be kept damp enough to result in condensation forming with the night time drop in temperature. A light spraying will ensure the humidity is kept quite high (80%). *A. avicularia* is a popular species for both the beginner and the more experienced keeper. Handling is not recommended as it is reluctant to leave its web and it can also move quite fast. As long as a large enough container is used and adequate food allowed a number of *A. avicularia* can be kept together. The females will construct their web and not intrude on each other's territory. Mature males will wander in search of females.

The Pink Toed tarantula lives up to its name when adult. It is black apart from the pink 'toes'. The spiderlings are a negative image of the adults. They are basically pink with black chevron markings on the abdomen and the 'toes' are black. As they grow they attain their adult colours.

*Avicularia nigrotineata* is very similar to *A. avicularia*, the obvious difference being that the 'toes' are yellow, hence its common name Yellow Toed Tarantula. It requires the same conditions as *A. avicularia*. This is a good beginner's spider.

*Avicularia versicolor* (Walckenaer 1837) – The Martinique Red Tree Spider. This is a relatively new species to the hobby but it has already been successfully bred in Europe. It is found in the islands of Martinique and Guadeloupe where it lives in rain forests. It is an arboreal species so requires height rather than floor area. It drinks from the condensation that forms on its web and the sides of its container. To keep the humidity level high, a light spraying is necessary. As its name indicates, this spider changes its colour. This happens as it approaches maturity. Juvenile specimens are almost sky blue with some grey hairs. As the spider approaches maturity the colour changes from blue to reddish grey. The web construction of *A. versicolor* is similar to that of *A. avicularia* but the adult female is slightly larger. This species as with *A. avicularia* lives better if there is a relatively large volume of air around it, so spiderlings should not be kept in 5cm. × 5cm. pots recommended for most other species. They should be started in containers at least 12cm. in diameter and 5cm. deep.

*Grammostola spatulatus* (Cambridge 1897) – Chilean Rose or Chilean Common Tarantula. This is the ideal beginner's species, if somewhat uninteresting. They are very docile and rarely make an attempt to move quickly. They are quite hardy and do not require feeding often. *G. spatulatus* is an attractive spider. It is basically brown but its abdomen and legs are covered with pink hairs. The carapace has a pink sheen. Most of the specimens available for sale are imported and are relatively inexpensive. The species has been bred in captivity but the spiderlings are extremely slow growing, indicating a long life span. The spiderlings are pink with a dark grey/black abdomen. Due to the fact that this species is so docile, mating is quite straight forward as neither male nor female show aggression. The natural habitat of *G. spatulatus* is in Mediterranean type vegetation in Chile where it lives in shallow burrows. Humidity is around 75%. Although considered by many to be rather boring, *G. spatulatus* is ideal for the beginner and careful handling is not a problem.

It is ironical that this, one of the least aggressive species, was used in the film adaptation of Dennis Wheatley's black magic book, 'The Devil Rides Out'.

*Grammostola cala* (Chamberlin 1917) – The Chilean Beautiful. This species is almost identical to *G. spatulatus*. It originates from the same part of the world, but a different habitat. It is found in the coastal rain forests of

Chile, therefore requires humidity of 80%. It is a good speces for the beginner as it is docile and hardy. The colouring is the same as that of *G. spatulatus*, brown with pinkish hairs on the legs and abdomen and a pink sheen on the carapace. Unfortunately it is impossible to know which species is being bought when one is purchased in a pet shop and this makes breeding difficult. It is possible to differentiate between the species if the moult skin is used. The spermathecae are different. Once it has been established that male and female are of the same species, mating is straight forward but, as with *G. spatulatus*, the young are very slow to reach maturity.

*Grammostola pulchripes* (Simon 1891) - The Pampas Tawny Red. This is a small but attractive species. It is docile and quite hardy and a good beginner's spider. They are becoming more difficult to obtain. They have been bred in captivity but, as with all members of this genus, are very slow growing. Adult females are dark brown with tawny red hairs on the abdomen. In the centre of the dorsal side of the abdomen is a patch of golden hairs which reflect the light. This makes *G. pulchripes* particularly attractive. The males have similar colouration but have much longer legs. The spiderlings are pink with a dark grey/black abdomen. The natural habitat of *G. pulchripes* are the Pampas/savannah grassland areas of Brazil and Paraguay so humidity of 75% is required.

*Grammostola grandicola* (Strand 1908) - The Common Negra Grey. *G. grandicola* is a fairly uncommon species in the United Kingdom. Small numbers were imported a few years ago. It is rather small compared to other more popular species, so was not in such great demand. This is unfortunate as it is attractive and docile. The basic colour is black with dark grey hairs on the abdomen and a freshly moulted specimen can be quite striking. Its natural habitat is the mixed tropical scrub or savannah of Brazil, Argentina and Paraguay. If obtainable it is a good beginner's spider and every effort should be made to breed them in captivity if a male becomes available.

*Phryxotrichus auratus* (Pocock 1903) - The Chilean Yellow Rump. *P. auratus*, sometimes named *P. roseus* is very similar to *G. cala* and *G. spatulatus*. It originates from Chile and Peru where its habitat is mountain scrub. Humidity need be no more than 75%. It is similar in size, shape and temperament to its neighbours, but is somewhat darker in colour. There is a variation of colour, from yellowish hairs on the abdomen to foxy red hairs. This is also a very docile spider which will go for long periods without eating. It is extremely slow growing, indicating a long life, but if spiderlings are produced they will take perhaps 6 or 7 years to reach maturity. Males of this species, as well as *G. spatulatus* and *G. cala*, have been known to live over two years after reaching maturity and an adult female will have an inter-moult period of about two years.

*Ephebopus murinus* (Walckenaer 1837) – The Yellow Flame Knee Tarantula. This is a fairly new species to the hobby and originates from Brazil. It lives in silk lined tubes in sandy soil and requires about 75% humidity. The carapace is dark brown with short sandy yellow hairs. The abdomen is a rich chocolate brown with scattered reddish hairs, more numerous towards the posterior. The legs are the same rich chocolate brown, but there are two distinct yellow bands on the patella. These become longitudinal on the tibia with a single yellow line on the metatarsus. This can be an aggressive spider so handling is inadvisable.

*Pterinopelma saltator* (Pocock 1903) – The White Collared Tarantula. *P. saltator* is in great demand, but unfortunately is very rarely available. It is surely the most docile of all the tarantulas and easily handled, hence its popularity. It is quite a small species. The abdomen is dark brown and covered with dark brown and foxy red hairs, as are the legs. The legs are highlighted by pale transverse bands at the extremities of the leg segments. The common name is due to the distinct pale band across the base of the chelicerae which is said to resemble a vicar's collar. *P. saltator* originates from the pampas plains of Uruguay and Argentina where it burrows. Humidity should be about 75%. It will fast for quite long periods, up to two months, and then begin to take food again. It cannot be assumed that it is approaching a moult so food should be offered and removed if uneaten. Moults in adults occur at about two yearly intervals. This is a very slow growing therefore relatively long lived species. They have been bred in captivity but this is a very slow process. It was seven months after mating that the eggs were layed and a further two before the spiderlings emerged. It is thought that it will be five years or more before these reach maturity. Further matings have taken place. Although the life span of a female is long, a mature male only survives for a few months after his final moult. *P. saltator* may be overlooked as a desirable species to keep because of its size but it is ideal for the beginner and interesting for the more experienced.

*Theraphosa leblondi* (LaTreille 1804) – The Goliath Bird Eater. *Theraphosa leblondi* is the giant of the spider world. It has a place in the Guinness Book of Records and can attain a leg span of 25cm. As well as this massive leg span it is a heavy bodied tarantula which requires frequent feeding. Its natural habitat is the rain forests of Surinam, Brazil, Guyana and Venezuela where it is said to live in swampy areas. There is some question about this as the original captive specimens were kept in conditions of up to 100% humidity. Experience has shown that although high humidity is required it need not exceed 80%. Wild caught individuals only had their size to offer as an attraction as they were otherwise dull brown spiders which had a rather ragged appearance. After a moult they are magnificent creatures with a full compliment of hair which ranges from deep brown to almost black. It is not unduely aggressive but its urticating hairs cause an extremely itchy rash. It also has the ability to stridulate, that is to make a hissing sound. Which can be used to frighten off potential predators. Captive bred spiderlings grow fast and eat well. They soon outgrow their containers and when adult require a floor area of at least 45cm. × 30cm. To allow for the extra humidity a plastic container can be buried in the substrate, covering up to half of the area and this thoroughly soaked. Drinking water can be provided as well. *T. leblondi* is not a spider for a beginner, not least due to the cost. It is less tolerant of unsuitable conditions. Handling is inadvisable because of the irritating nature of the hairs.

*Pseudotheraphosa apophysis* (Tinter 1991) - The Pink Footed Goliath. This species was initially imported into the United Kingdom as *Theraphosa leblondi*. One of the adults imported as such produced spiderlings which were nothing like the young of *T. leblondi*. The adults are very similar, but the difference in the spiderlings is that the last two segments of the legs are pink. The spiderlings of *T. leblondi* are completely brown. It can only be assumed that the pink markings will disappear as the spider matures. The imported specimens came from Venezuela, and only scientific investigation of dead specimens will identify them correctly. *P. apophysis* requires similar living conditions to *T. leblondi*, coming from tropical rain forest. Not only is this a new species, but a new genus as well.

*Lasiodora parahybana* (Mello-Leitao 1917) - The Campina Grande Salmon Pink Bird Eater. *L. parahybana* is second only in size to *Theraphosa leblondi*. It is not such a heavily built spider but larger leg spans have been recorded, particularly in mature males. This is basically a brown tarantula, but adults have pinkish hairs on the abdomen and legs. The labium is salmon pink. The majority of specimens available are captive bred and the females take about three years to reach maturity. It is not a particularly aggressive spider but is very fast moving so is not recommended for handling. The natural habitat is the rain forests of Brazil so humidity of 80% is needed. This is a relatively fast growing tarantula and a freshly moulted adult is very striking. They are not difficult to keep and breed in captivity quite reaily, more success having been had with these than *T. leblondi* to date.

*Pamphobeteus fortis* (Ausserer 1875) - The Giant Columbian Bird Eater. *P. antinous* (Pocock 1903) - The Bolivian Steely Blue Legged Bird Eater. *P.*

*insignis* (Pocock 1903) – The Columbian Purple Bloom Bird Eater. *P. roseus* (Mello-Leitao 1923) – The Giant Pink Starburst Bird Eater. *Pamphobeteus* species have been grouped together because many have been imported into Britain and on occasions identification has been doubtful. They originate from the rain forests of Peru, Columbia and Bolivia and require 80% humidity. They are all large, heavy bodied spiders which can show aggression if provoked. They are voracious eaters in the inter-moult period and the spiderlings grow quite fast. Although quite aggressive they are extremely gentle with each other during courtship and mating. During the period after a moult the attractive colours fade to an almost uniform brown, hence the difficulty in identifying living, wild caught specimens. *P. fortis* can be almost black on moulting with a pinkish sheen on the carapace. In a mature male the carapace is quite striking. *P. antinous* is blackish brown with a striking steel blue lustre on the legs in natural light. *P. insignis* has deep olive brown body hairs with a blueish purple 'bloom' on the limbs. *P. roseus* has a black abdomen with long pinkish hairs. The carapace is pink. These giant bird eaters require a little more space than average but are not difficult to keep. They are better suited to those with some experience. Some tarantulas have been imported from the same areas of south America reputedly of the genera *Acanthoscuria* and *Xenesthis* as these are similar in size and build to *Pamphobeteus*. This makes breeding difficult, so correct identification should be ensured before attempting this.

# African Species

*Ceratogyrus darlingi* (Pocock 1897) – The East African Horned Baboon. *C. darlingi* is unusual in that instead of a groove in the foveal area of the carapace, it has a horn projecting towards the rear. It is a fawny brown spider with an orange band across the abdomen. The carapace is dark with radiating white lines. The natural habitat of *C. darlingi* is deep burrows in Zimbabwe and Mozambique, in dry forest and grassland. The burrows are to escape from the severe heat and drought of the dry season. These spiders do not require high humidity, but on the other hand should not be allowed to dry out completely. It was observed that in the wild the spiderlings emerge at the beginning of the dry season. Many captive bred spiderlings perished by being kept in too humid an atmosphere. The substrate should be barely damp, much drier than for the majority of species. *C. darlingi* has the aggressive nature of baboon spiders and should not be handled. It can move extremely fast.

*Ceratogyrus bechuanicus* (Purcell 1902) – The Starburst Horned Baboon. Along with other species of the Ceratogyrus genus, *C. bechuanicus* has a foveal horn pointing towards the rear. It lives in deep burrows and is found in Botswana (formerly Bechuanaland) and northern South Africa. The same criteria for keeping and breeding should be observed as for *C. darlingi*. The two species are very similar, *C. bechuanicus* being darker in colour. It does appear to be more aggressive than *C. darlingi* and will not tolerate the presence of the male except during mating.

*Stromatopelma calceata* (Fabricius 1793) – The Feather-leg Baboon. *S. calceata* is a typical baboon spider – fast and aggressive. It is a very attractive and interesting species to keep for the more experienced. There were two species known in the past as *Scodra calceata* and *S. grisepes*. The genetic name was changed because it had earlier been given to a butterfly genus. Recent investigations have shown that the two species are synonymous, being too similar to warrant a species each. Most of the specimens available are *Stromatopelma calceata*, but those thought to be *S. grisepes* are *S. calceata grisepes*, a sub-species. The species is found over a wide range of West and Equatorial Africa around the Gulf of Guinea, ranging from Guinea to the Cameroons. Many captive specimens originated from the Ivory Coast. This is an arboreal species, making its tube web at the base of palm fronds high above the ground, both in the lower forest regions and the higher grass/scrubland areas. The overall colour of this spider is fawny brown with a starburst pattern on the carapace. The abdomen has an irregular pattern of fawn and dark greyish brown. The legs are fawn with a black spot on the tarsus and metatarsus. The spiderlings grow quickly and the males will mature in a year. Breeding has been successful despite the fact the female has a propensity to devour the male as soon as he has mated. The egg-sac is constructed on and attached to the web and the female guards it ferociously. Care must be taken when removing the spiderlings after their second moult. The female should be removed before this is attempted. *S. calceata* has the typical posture of an arboreal spider, lying flat with two pairs of legs pointing forwards and two back. Arboreal conditions should be provided with great care being taken that the spider cannot escape.

*Citharischius crawshayi* (Pocock 1900) – The King Baboon or Kinani Rusty Red Baboon. *C. crawshayi* deserves the common name King Baboon. It is an extremely impressive African spider, being only slightly smaller than the largest from that continent, *Hysterocrates hercules* from Upper Niger. *C. crawshayi* is found in Kenya, Uganda and Tanzania where it lives in burrows in acacia scrub and grassland. The burrows are about 50cm. deep with a chamber at the end and are angled among the roots of acacia bushes. The countries of origin have distinct dry and rainy seasons so these spiders can regulate the humidity by retreating to their burrows when it is very dry. *C. crawshayi* is a beautiful reddish brown with rusty red hairs. It has the appearance of being covered with velvet. The legs are particularly impressive, especially the fourth pair. They have a thick covering of rusty red hairs, giving them a sturdier appearance. The tarsi on the rear legs are turned inwards, it is thought to facilitate burrowing. An interesting feature of this tarantula is that it stridulates. When disturbed it can produce quite a loud hissing sound which it keeps up for quite a while. This is a defense which undoubtedly frightens off predators. *C. crawshayi* is not overly aggressive but it does not like to be disturbed. It is not recommended for the beginner, least of all because of the high cost. There have been spiderlings available in Britain but these have been produced by gravid wild caught females. The male of this species is so unlike the female that it was originally classified as a different genus and species. The colouring is similar, but the hairs on the legs are much longer and it is a lot smaller. Hopefully males will mature from the spiderlings quite soon.

*Pterinochilus murinus* (Pocock 1897) – The Mombasa Golden Starburst Baboon. This is one of the most aggressive of the Baboon spiders. It will readily rear up with fangs bared in a typical threat posture. This is particularly so if a female is guarding an egg-sac. These spiders should be treated with respect and never handled. A number of people have been bitten, the effects ranging from discomfort at the site of the puncture wound to hospitalisation. As with all spider venom the sensitivity of the recipient has much to do with the reaction, so avoiding being bitten is the best solution. The threat posture is a warning. *P. murinus* is an attractively marked tarantula. The abdomen is mustard yellow with a dark grey pattern. The carapace is black with a golden yellow starburst pattern. The male is less strikingly marked and smaller than a mature female. Despite their aggressive nature towards humans the male and female will live together until the egg-sac is produced when the male should be removed. Many of the imported females were gravid, resulting in a glut of spiderlings. A female can produce up to three egg-sacs at intervals of about two months, all the result of a single mating. The species is relatively short lived, a male maturing in a year, a female living about four years. The natural habitat of *P. murinus* is in Kenya and Tanzania where it is an opportunist. It excavates a chamber under a rock or log and lines it with silk. It will also utilise an uninhabited lizard burrow. The species is common in populated areas where it lives under raised floors feeding on insects attracted by human habitation. It produces a lot of silk to line its retreat and the females attach the egg-sac to this.

*Pterinochilus vorax* (Strand 1907) – The Lake Tanganyikan Golden Baboon. This is an aggressive species which should not be handled. It is widely distributed in western Tanzania and eastern Zaire where its natural habitat is dry grassland and woodland. It makes a retreat of web but as yet there is no data as to how it lives in the wild. Twigs and bark should be provided to which it can attach its web. It is not such a copious producer of silk as *P. murinus*, neither is it quite so aggressive. Also it is a much sturdier spider. The abdomen is dark fawny brown with indistinct chevron markings. The first two pairs of legs have bright orange hairs and there is an orange stripe across the underside of the abdomen. The spiderlings of this species were observed to be attracted by sunlight where they congregated before being separated.

*Pterinochilus sjostedti* (Tullgren 1910) – The Kilimanjaro Baboon. *P. sjostedti*, as its common name indicates, originates from the area around

Mt. Kilimanjaro, in Kenya and Tanzania. Its natural habitat is acacia scrubland and it is reputed to have been found at an altitude of over 1000 meters. There is great colour variation within this species, but those available are generally a fairly light greyish brown with white markings on the leg joints. The abdomen has typical chevron markings. This is not a particularly aggressive baboon spider, but handling should be avoided. It is a relatively new species to be imported, but captive bred spiderlings have been produced.

*Pterinochilus affinis* (Tullgren 1910) - The Highland baboon. *P. affinis* shares the same natural habitat as *P. sjostedti*, the Mt. Kilimanjaro region of Kenya and Tanzania. The natural vegetation is grassland ans acacia scrubland. They are both of a similar build but *P. affinis* is much darker in colour. It is sometimes called the Kilimanjaro Sooty Black Baboon. The original imported specimens were extremely aggressive, but captive reared spiderlings (many females were gravid) do not show this trait. However care should be taken as all baboon spiders are unpredictable and should not be handled. As with all East African species, humidity should not be too high, around 65% being adequate, and the container should have the substrate damp at one end only. Provision should be made for *P. affinis* to attach its web.

*Ischnocolus hancocki* (Smith 1990) - The Larache Gold. *I. hancocki* is a small species found in Morocco. It is rather similar in appearance to the northern European species of araneomorph, *Drassodes lapidosus*, commonly known as the mouse spider. The hairs on the abdomens of both these species resemble the silky fur on a mouse's body. Closer examination of the underside shows two pairs of book-lungs and downward striking fangs indicating that *I. hancocki* is a theraphosid spider. The colour is light olive brown with golden yellow hairs. The abdomen has a median line with chevrons. The collection site of the type specimen was a rocky hillside in north west Morocco on the site of ancient ruins. The spiders were found under rocks, the young not having begun a burrow. The mature specimens were found in burrows, under rocks, having lined the burrow and the cavity with silk. The burrows were 15cm. to 20cm. deep. The vegetation of the area was low and sparse, being heavily grazed, but the rock strewn hillside provided plenty of retreats. It was an area of Mediterranean type climate which has low rainfall and average temperatures. *I. hancocki* does not require a lot of space, but a retreat can be made of cork bark. The spiderlings readily burrow in Vermiculite. This species was discovered in 1988 by the authors and has since been bred by them.

*Chaetopelma olivaceum* (Ausserer 1871) - The Middle East Olive Gold.

Although this species has been included with those from Africa, its distribution is centred in Syria and Lebanon. It is also found in Egypt and is widely dispersed throughout the Middle East. It is quite a small tarantula but it is relatively active. It is a dark olive brown with short hairs on the abdomen. It is an opportunist as far as burrows are concerned, excavating a chamber below a rock and lining it with silk. As its natural habitat is dry, low humidity is required and average temperature. The spiderlings mature in around two years and this species has been bred in captivity.

# Asian Species

*Haplopelma minax* (Simon 1886) – The Thailand Black or Edible Tarantula. As with all Asian species of tarantula, *H. minax* is aggressive. It is quick to assume a threat posture with legs raised and fangs bared. In fact it often falls onto its back where it continues to threaten after the danger (i.e. the human) has passed. Despite this behaviour it is reluctant to actually attack, but should nonetheless never be handled and always be treated with respect. An adult female has a dark grey abdomen with black chevron markings. The legs are black with foxy red/brown hairs on the inside faces. These hairs darken with successive moults after maturity. The males, when mature, are distinctively different. Before the final maturing moult they are outwardly identical in colouring to sub-adult and young adult females. The final moult reveals a fawny brown spider with black tarsi. The male is usually much smaller than the female. Mating is quite a peaceful affair if the female is receptive. Although the female makes a considerable amount of silk, she carries and turns her egg-sac unlike some African Baboon spiders. The spiderlings are easy to rear, reaching maturity in 3 to 4 years for a female. *H. minax* is a burrowing species which is found in the rain forests of Burma and Thailand. High humidity is needed (80%). In their native lands these spiders are considered a delicacy and are barbecued on skewers by the natives.

The generic names *Haplopelma* and *Melopoeus* are synonymous so this species may be called *H. minax* or *M. minax*.

*Haplopelma albostriatus* (Thorell 1897) – The Thai Tiger or Lesser Thailand Black. The abdominal markings of *H. albostriatus* are similar to those of *H. minax*. The legs, however are unmistakable. They have longitudinal cream lines similar to those of the Costa Rican Zebra. It is a smaller species than *H. minax*, but conditions for keeping it are the same. It is a burrowing species from Thailand and Burma where it lives in rain forest and bamboo groves. Humidity should be around 80%. This is an aggressive species which should not be handled. Again the genetic name may be given as *Melopoeus*.

*Haplopelma lividus* – The Burmese Blue. *H. lividus* is a very striking spider. Its abdomen is almost black with slight chevron markings and its carapace is dark beige. A freshly moulted specimen has vivid blue hairs on its legs when they catch the light. The mature male is rather insignificant having more obvious chevrons on its abdomen and black legs. It is considerably smaller than the adult female. Attempts to breed this species in captivity have, as yet, been unsuccessful due to the aggressive nature of the female. Males are immediately pounced upon and eaten. The only spiderlings that have been available have been from gravid wild caught females. *H. lividus* constructs a lot of web and this is a golden colour. It originates from the rain forest areas of Burma where it lives in burrows. Humidity needs to be high (80%). This is not a suitable spider for a beginner and should never be handled.

*Coremiocnemus validus* (Pocock 1895) – The Malaysian Feather Leg or Singapore Brown. This is a rather uninteresting species, at first glance. It is reddish brown in colour and during the hours of daylight it rarely moves but has a hunched up appearance, with its 'knees' almost meeting above its body. It is a nocturnal species and when it begins to move demonstrates a huge leg span compared to its body size. The fourth pair of legs have quite long hairs giving them a feather like appearance. The male is considerably smaller than the female and is fawn in colour except for the femur on each leg which is black. *C. validus* is found in the rain forest areas of Malaya, Singapore and, it is thought, in India. It is intolerant of less than ideal conditions which should have high humidity (80%) and a temperature about 26°C.

*Poecilotheria regalis* (Pocock 1899) – The Indian Ornamental. The *Poecilotheria* genus contains some of the most beautifully marked tarantulas. *P. regalis*, the logo of the British Tarantula Society, was the first species of this genus to become available in the United Kingdom. They are arboreal spiders which live behind bark and in hollow trees in the forest areas of southern India and Sri Lanka. The markings aid camouflage. The carapace is grey with a dark longitudinal stripe which is broken up by lighter markings. The abdomen is dark with a broad white longitudinal band which is edged in black. Stripes radiate out from this. The legs are marked with black, white and grey patches. The underside of the abdomen has a broad yellow band at the anterior end and the underside of the anterior legs is also marked with yellow. This will be visible in a threat posture. Humidity of 75% to 80% is required by these spiders and a temperature around 25°C. A piece of bark should be placed in the container behind which they can construct a web. This is a very attractive spider which has now been bred successfully in captivity. The spiderlings are fairly hardy and mature quickly.

*Poecilotheria subfusca* (Pocock 1895) – The Sri Lankan Ornamental Ivory Spot. This species resembles *P. regalis* in shape, habits and markings, although the colouration is more brown. It has been found living in hollow trees in Sri Lanka where a number had congregated together. This was probably to lessen the temperature drop, as the area in which they are found is quite cool at night. It is not recommended that they are kept communally in captivity. Their requirements are the same as those for *P. regalis*. This species is not widely available in the United Kingdom yet.

*Poecilotheria fasciata* (Latreille 1804) – The Sri Lankan Ornamental Black and White. *Poecilotheria striata* (Pocock 1895) – The Mysore Orange Legs. Both these species have been confused with *P. regalis*. There is only a slight variation in markings between *P. fasciata* and *P. regalis*, so without positive identification mating should not be attempted. *P. striata* lacks the pale band on the underside of the abdomen. Also the yellow markings on the anterior legs of *P. regalis* are more orange on *P. striata*.

Tarantulas of the *Poecilotheria* genus are much sought after. They all need the same basic conditions as they are all arboreal species and come from similar habitats.

*Chilobrachys sericeus* (Thorell 1895) – The Rangoon Mustard Tarantula. *C. sericeus* is one of the larger Asian tarantulas, the female growing to quite an impressive size. They come from the rain forest areas of Burma so need high humidity (80%). An adult female is dark brown with mustard coloured hairs on the abdomen. They are not particularly aggressive, but retreat into their web when disturbed. They have been bred in captivity despite the diminutive size of the male. He is lighter in colour than the female. At only a fraction of her size he mates repeatedly with her before retreating and, although it did not seem possible, fertile egg-sacs have been produced. The female carries her egg-sac until the spiderlings emerge.

*Selenocosmia lyra* (Strand 1913) *Selenocosmia javanensis* (Walckenaer 1837) Many species of this genus have been imported into the United Kingdom without having been identified. Usually they are sold as *Selenocosmia sp.* which makes breeding very difficult if the production of hybrids is to be avoided. They are all rather aggressive spiders many of which come from the East Indian Islands – Java, Sumatra, New Guinea and Borneo as well as the Asian continent – Burma, Pakistan and India. They are copious web builders. Most species are of a uniform colour, either black or brown, with short hairs on the abdomen. One species which has been positively identified is *S. lyra*. A mature female is dark brown, almost black. A mature male is somewhat lighter in colour with longer hairs. He is about the same size as the female. The species originates from the rain forests of Sumatra. *S. lyra* is known as the Banda Kwala Dark Brown Tarantula. *S. javanensis*, the Java Yellow Kneed Tarantula has also been imported and identified. It is basically a brown spider. To date it has not been bred in captivity. Some tarantulas belonging to the *Selenocosmia* genus are quite large, but are not very popular due possibly to the fact that they are not very colourful. They are aggressive and make a lot of web.

# GLOSSARY

**Accessory gland** – secretes a liquid which is mixed with the sperm.

**Apolysis** – separation of the old cuticle from the hyperdermal cells.

**Araneidae** – a major family of orb-weaving spiders.

**Araneomorphae** – true spiders (Labidognatha).

**Atrax** – the genus which includes the Sydney Funnel Web, *Atrax robustus*.

**Atypidae** – the Purse Web spiders.

**Avicularia** – genus of Theraphosidae which includes the bird eating spiders.

**Book-lung** – the 'leaves' increase the area through which oxygen can be absorbed.

**Carapace** – the hard shield covering the cephalothorax.

**Cephalothorax** – the fused head and thorax of spiders.

**Chelicerae** – the jaws of spiders.

**Clypeus** – the area between the anterior eyes and the edge of the carapace.

**Ctenidae** – the Wandering spiders.

**Ctenizidae** – the Trap Door spiders.

**Dipluridae** – Funnel web spiders with long spinnerets.

**Ecdysis** – the shedding of the old exoskeleton; moulting.

**Embolus** – the tip of the male apparatus for injecting sperm.

**Epigastric furrow** – a fold in the ventral skin of the abdomen, between the anterior book-lungs.

**Epigyne** – structure on the underside of the abdomen to produce or receive sperm.

**Exuvial space** – a space created between the old and the new developing exoskeleton.

**Fang** – the piercing part of the spider's jaw producing poison.

**Fovea** – a depression in the carapace.

**Gastrocantha** – orb weavers with brightly coloured spiney abdomens.

**Gonopore** – opening to reproductive organs.

**Ischonothele** – a South American and African Dipluridae.

**Labidognatha** – synonym of Araneomorphae.

**Labium** – a plate which forms the back of the spider's mouth.

**Latrodectus** – the Widow spiders.

**Loxoscelidae** – the Violin spiders.

**Lycosidae** – the Wolf spiders.

**Maxillae** – a pair of accessory jaws to break up the food.

**Mesothelae** – a primitive form of spider which is segmented.

**Mygalomorphae** – orthognatha whose fangs strike downwards.

**Nephilinae** – the Golden Silk spiders.

**Opisthosoma** – the abdomen.

**Palp** – adapted in the male as sexual organs.

**Pedicel** – narrow waist connecting the prosoma and opisthosoma.

**Pholcus** – the Daddy-long legs spiders.

**Prosoma** – cephalothorax.

**Rastellum** – a row of teeth on some Mygalomorph spiders, used for digging a burrow.

**Rostral plate** – a filtration 'plate' within the pharynx.

**Salticidae** – the Jumping spiders.

**Spermathacea** – a sac within the female's abdomen in which sperm is stored.

**Spermatophore** – what the male scorpion secretes and the female absorbs during mating.

**Spinneret** – the structure through which silk is produced.

**Stabilimentum** – a band of silk placed across the orb web of certain spiders.

**Sternum** – the plate of cuticle that forms the underside of the cephalothorax.

**Tarantula** – a common name for Theraphosidae.

**Trachea** – a tube leading to the atmosphere into the spider which is used for breathing.

**Urticating hairs** – hairs thrown from the abdomen of a tarantula as defense.

# REFERENCES AND FURTHER READING

Bristowe, W.S. 1941. The Comity of Spiders. London: The Ray Society.
Bristowe, W.S. 1958. The World of Spiders. London: Collins.
Comstock, J.H. 1980. The Spider Book. London: Cornell University Press.
Crompton, John. 1950. The Spider. London: Collins.
Daniken, Erich Von. 1974. In Search of Ancient Gods. Souvenir Press Ltd.
Foelix, Rainer F. 1982. Biology of Spiders. Cambridge, Massachusetts: Harvard University Press.
Hancock, Kathleen and John. 1989. Sex Determination of Immature Theraphosids from their Cast Skins. Southminster, Essex: K. and J. Hancock.
Hancock, Kathleen and John. 1990. Simply Tarantulas – A Guide for the Beginner. Southminster, Essex: K. and J. Hancock.
Hillyard, P.D. and Sankey, J.H.P. 1989. Harvestmen. The Linnean Society of London.
Jones, Dick. 1983. Spiders of Britain and Northern Europe. Feltham, Middlesex: Country Life Books.
Jones, Dick. 1986. Spider. London: Orbis Book Publishing.
Levi, Herbert W. Spiders and their Kin. New York: Golden Press.
Murphy, Frances. 1980. Keeping Spiders, Insects and other Land Invertebrates in Captivity. Edinburgh: Bartholomew.
Preston-Mafham, Rod and Ken. 1984. Spiders of the World. Poole, Dorset: Blandford Press.
Raven, Robert J. 1985. The Spider Infraorder Mygalomorphae (Araneae): Cladistics and Systematics. New York: Bulletin of the American Museum of Natural History.
Roberts, Michael J. 1885. The Spiders of Great Britain and Ireland. Vols 1-3 Colchester, Essex: Harley Books.

Savory, Theodore. 1977. Arachnida. London: The Academic Press.
Savory, Theodore. 1971. Spiders. London: Ginn.
Shear, William A. 1986. Spiders, Webs Behavior and Evolution. Stanford University Press.
Smith, Andrew. 1986. The Tarantula Classification and Identification Guide. London: Fitzgerald Publishing.
Smith, Andrew. 1990. Baboon Spiders - Tarantulas of Africa and the Middle East. London: Fitzgerald Publishing.
Snow, Keith R. 1970 The Arachnids - an Introduction. London: Routledge and Kegan Paul.

# INDEX

abdomen, 9, 12, 13, 53, 62, 79
*Acanthoscuria sp.*, 100
Acari, 106
accessory glands, 64
alcove, 62
Amblypygi, 101
anatomy, 9, 58
*Androctonus australis*, 100
*Anidiops villosus*, 78
anterior prosoma, 9
aorta, 12
*Aphonopelma chalcodes*, 120
*A. seemanni*, 119-120
apolysis, 53
aquarium, 27, 34, 72
Arachne, 1
Arachnida, 5
arachnids, 41
arachnology, 1
arachnophobia, 20
Araneae, 5, 15
Araneidae, 81, 88
Araneomorphae, 6, 15, 81
*Araneus diadematus*, 89
arboreal species, 24, 29, 32, 35, 37, 40, 41, 48, 71, 120, 123, 124, 135, 136
*Argiope bruennichi*, 88
arthropods, 9
Asian Black Scorpion, 97
Athena, 1
*Atrax robustus*, 7, 12, 77
Atypidae, 6, 15, 75

*Atypus*
- *affinis*, 6, 75, 80
- *bicolor*, 6
autotomy, 57
*Avicularia*
- *avicularia*, 28, 123
- *nigrotineata*, 124
- *versicolor*, 124
Aviculariinae, 15

Baboon Spiders, 3, 129-132
"balloon", 77
Banda Kwala Dark Brown Tarantula, 137
Bird Eaters, 123, 126-128
Black Velvet, 118
Blonde, 120
book-lungs, 6, 9, 12, 34, 59, 62, 75, 81
breeding, 67-70
Burmese Blue, 36, 135
burrow, 24-26, 37, 75, 77-80
burrowing species, 24
Buthidae, 93, 98
*Buthus occitanus*, 100

Camel Spiders, 100
Campina Grande Salmon Pink, 127
carapace, 9, 12, 16, 54
caterpillars, 51
central nervous system, 11
*Centruroides sp.*, 100
cephalothorax, 9, 75

*Ceratogyrus*
- *bechuanicus*, 34, 129
- *darlingi*, 12, 129
*Chaetopelma olivaceum*, 132–133
chelicerae, 7, 9, 10, 11, 18, 43, 54, 75, 78
Chilean Beautiful, 124–125
Chilean Rose, 21, 25, 28, 30, 69, 124
Chilean Yellow Rump, 125–126
*Chilobrachys sericeus*, 136
CITES, 111, 117
*Citharacanthus crinirufus*, 122
*Citharischius crawshayi*, 130–131
classification, 14
cockroaches, 50
coconut fibre, 36
collecting, 111–116
Common Negra Grey, 125
conservation, 111–116
containers, 24–29
*Coremiocnemus validus*, 135
cork bark, 25, 29, 40
coxa, 9, 13
Costa Rican Blue Front, 122
Costa Rican Red Leg, 119
Costa Rican Sun Tiger Abdomen, 121–122
Costa Rican Zebra, 119–120
crickets, 42, 43, 46–48
Ctenidae, 82, 89, 90
Ctenizidae, 6, 11, 15, 75, 78, 80
*Cupiennius salei*, 90, 91
Curly Hair, 117
cuticle, 53, 60
*Cyclocosmia sp.*, 79
*Cyclosternum fasciatus*, 121–122

Dangerous Wild Animals Licence, 81, 93
dead food, 52
Dipluridae, 6, 7, 12, 15, 75, 77, 78
*Dolomedes*
- *fimbriatus*, 85
- *plantarius*, 85
drinking water, 35
*Drosophila sp.*, 48–49
*Dugesiella epicureana*, 122

earthworms, 51

East African Horned Baboon, 129
ecdysis, 53–54
Edible Tarantula, 134
egg sacs, 11, 13, 68–70, 77, 78
embolus, 18, 62
Emperor Scorpion, 93, 96
*Ephebopus murinus*, 126
epigastric furrow, 12, 58–60, 62
epigyne, 18, 68
*Euathlus*
- *albopilosa*, 117
- *emilia*, 14, 118, 119
- *mesomelas*, 14, 119
- *smithi*, 7, 14, 15, 25, 30, 111, 117, 118
- *vagans*, 14, 118
eurypterids, 1
Eumenophorinae, 15
*Euscprpius flavicaudis*, 94, 97
exoskeleton, 53
exuvial space, 53
eyes, 9, 16

False Scorpions, 106
fangs, 6, 9–11, 43, 54, 68, 75, 78
Fat Tailed Scorpions, 100
Featherleg Baboon, 129–130
feeding, 41–52
feeding hole, 29
femur, 9
Flame Knee, 118
flies, 48–49
forest bark, 36, 37
fovea, 12
foveal groove, 16
fruitflies, 48–50, 73
Funnel Web Spiders, 15, 75, 77

Garden Cross Spider, 89
*Gasteracantha sp.*, 89
genital organs, 18
genus, 14
Giant Huntsman Spider, 89
Golden Cuban Brown, 121
Goliath Bird Eater, 34, 126–127
gonopore, 59, 60, 62, 64
Gonyleptidae, 106
*Grammostola*
- *cala*, 124–125

– *grandicola*, 125
– *pulchripes*, 125
– *spatulatus*, 21, 25, 28, 30, 69, 124
Grammostolinae, 15
gravel, 36

Hairs, urticating, 7
Haitian Brown, 121
*Hapalopus incei*, 121
*Haplopelma*
– *albostriatus*, 14, 134-135
– *lividus*, 36, 135
– *minax*, 28, 69, 134
Harvestmen, 104
head, 9
heating, 30
*Heterometrus spinifer*, 97
*Heteropoda venatoria*, 89
*Heteroscodra crassipes*, 3
Highland Baboon, 132
hoppers, 44
Horned Baboon, 129
humidity, 26, 28, 29, 34, 35, 40, 73, 78, 80
hygiene, 29
*Hysterocrates hercules*, 130

identification, 16-18
Indian Ornamental, 40, 135-136
intestinal tract, 11
Ischnocolinae, 15
*Ischnocolus hancocki*, 132
*Ischnothele guyanensis*, 78
Israeli Gold Scorpion, 93, 100

Java Yellow Kneed Tarantula, 137
joint membrane, 9
Jumping Spiders, 92

Kilimanjaro Baboon, 131-132
Kilimanjaro Sooty Black Baboon, 132
Kinani Rusty Red Baboon, 130-131
King Baboon, 130-131

Labidognatha, 6, 15
labium, 9
Lake Tanganyikan Golden Baboon, 131
Laniotores, 106

Larache Gold, 132
Large Clawed Yellow Scorpion, 97
*Lasiodora parahybana*, 127
*Latrodectus*
– *geometricus*, 83
– *hesperus*, 83
– *mactans*, 82
legs, 9, 16
*Leiurus quinquestriatus*, 93, 100
Lesser Thailand Black, 134-135
*Liochelis australasiae*, 98
live food, 41-52
locusts, 43-45
*Loxosceles sp.*, 82
*Lycosa tarentula*, 81
Lycosidae, 81, 85

*Macrothele calpeiana*, 7, 12, 77, 78
maggots, 48-49, 73
Malaysian Feather Leg, 135
Martinique Red Tree Spider, 124
mating, 68-69
mating hooks, 58, 68
maturity, 58
maxillae, 18
mealworms, 51
*Melopoeus sp.*, 134-135
Mesothelae, 5, 12, 15
metatarsus, 9, 10
*Metriopelma zebratus*, 121-122
Mexican Red Knee, 7, 25, 30, 111, 117-118
microcrickets, 48, 73
microscope, 60
Middle East Olive Gold, 132-133
mites, 35, 36, 43, 106
Mombasa Golden Starburst Baboon, 131
moths, 51
moult, 9, 41
moulting, 40, 42, 46, 53-57, 68, 72
moult skin, 18, 60
muscles, 11, 12
Mygalomorph, 6, 15, 75, 78
Mysore Orange Legs, 136

Nazca lines, 2
*Nephila*

– *madagaskariensis*, 86
– *senegalensis*, 86
– *sp.*, 81
Nephilinae, 86,
Nursery Web Spiders, 86

Olive Brown Trinidad Spider, 121
oocytes, 59
Opiliones, 104
*Opisthocanthus crysopus*, 98
opisthosoma, 9, 12, 54
opportunists, 24
Orb Weavers, 86–89
Ornithoctoniae, 15
Orthognatha, 6, 15
ovaries, 12, 59, 69
oviducts, 59

Painted Tarantula, 118–119
Palomino, 120
palps, 68
Pampas Tawny Red, 125
*Pamphobeteus*
– *antinous*, 127–128
– *fortis*, 127–128
– *insignis*, 128
– *roseus*, 128
*Pandinus imperator*, 93–96
parasites, 21
parasitic wasps, 80
*Pardosa amentata*, 85
patella, 9
peat, 36, 37
pedicel, 12, 54
pedipalps, 9, 11, 43, 68, 93
Phasmids, 50
Pholcidae, 92
*Phoneutria fera*, 82, 91
*Phormictopus*
– *cancerides*, 121
– *nesiotes*, 121
*Phryxotrichus*
– *auratus*, 125–126
– *roseus*, 125
Pink Footed Goliath, 127
Pink Toed Tarantula, 28, 123
*Pisaura mirabilis*, 86
Pisauridae, 85

plants, 24, 25, 40
plastic containers, 28
*Poecilotheria*
– *fasciata*, 136
– *regalis*, 40, 135–136
– *striata*, 136
– *subfusca*, 136
poison, 11, 82
poison glands, 11
pooter, 46, 49, 50, 71, 73
*Portia africanus*, 92
posterior opisthosoma, 9
prosoma, 9–12, 54
*Psalmopoeus cambridgei*, 31, 120
Pseudoscorpiones, 106
*Pseudotheraphosa apophysis*, 127
*Pterinochilus*
– *affinis*, 132
– *murinus*, 131
– *sjostedti*, 131–132
– *vorax*, 131
*Pterinopelma saltator*, 126
pupae, 49
Purse Web Spiders, 6, 15, 75, 77

Raft Spiders, 85
Rangoon Mustard, 136
rastellum, 7, 78
records, 41
Red Back, 83
Red Rump, 118
regeneration, 57
relative humidity, 35
reproductive organs, 9, 12
retreat, 37, 40, 41, 77
*Rhytidicolus structor*, 79
rostrum plate, 43

Salticidae, 92
*Salticus scenicus*, 92
sand, 36, 37
scaffold web, 83
*Scodra*
– *calceata*, 129
– *grisepes*, 129
scopulae, 10, 16
*Scorpio maurus*, 97
Scorpions, 93–100

*Scytodes thoracica*, 92
*Selenocosmia*
- *javanensis*, 136-137
- *lyra*, 136-137
Selenocosmiinae, 15
sensory hairs, 10
setae, 78
sex determination, 58-66
*Sicarius sp.*, 82
Singapore Brown, 135
slit sensilla, 62
Solifugae, 110
Sparassidae, 82, 89
species, 14
sperm, 59, 68, 69
spermathecae, 18, 59, 60, 62, 64, 68
spermatophore, 96, 104
sperm web, 13, 68, 69
sphagnum moss, 40
spiderling, 29, 35-37, 54, 58, 60, 70-74, 77, 78, 80
spinnerets, 9, 12, 75, 77
Spitting Spider, 92
Sri Lankan Ornamental Black and White, 136
Sri Lankan Ornamental Ivory Spot, 136
stabilimentum, 89
Starburst Horned Baboon, 34, 129
*Steatoda paykulliana*, 83
sternum, 9
stick insects, 50
*Stothis astuta*, 79
stridulating organ, 16, 18
*Stromatopelma*
- *calceata*, 129-130
- *calceata grisepes*, 130
substrate, 25, 29, 31, 32, 34, 35, 71
sucking stomach, 11
Sun Spiders, 110
sweep net, 78
Sydney Funnel Web, 7, 77

Tail-less Whip Scorpions, 101
tarsus, 9, 10
*Tegenaria*
- *domestica*, 92
- *gigantea*, 92

temperature, 26, 72, 78
Thailand Black, 28, 69, 134
Thai Tiger, 134-135
*Thelechoris karschi*, 78
*Theraphosa leblondi*, 34, 126-127
Theraphosidae, 6, 7, 14, 15
Theraphosinae, 15
Theridiidae, 83
thermometer, 30, 34
thermostat, 30, 31, 34
thorax, 9
tibia, 9
Tiger Rump Doppel Ganger, 121-122
Tiny Malaysian Brown Scorpion, 98
trachea, 15, 34, 81
Trap Door Spiders, 6, 75, 78
Trinidad Chevron, 31, 120
trochanter, 9, 13
True Red Leg, 118-119
True Spiders, 6, 15, 81

Uropygi, 104
urticating hairs, 7
uterus externus, 59, 60, 62, 64

Velvet Mite, 106
venom, 13, 43, 78, 82, 83, 93
Vermiculite, 29, 35-37
vertebrates, 52

Wandering Spiders, 89
watering, 34
web, 22, 29, 37, 40, 69, 77, 78, 83
Whip Scorpions, 104
White Collared Tarantula, 126
Widows, 81-83
Widow, Black, 82
Widow, Brown, 83
Widow, False, 83
Wind Spiders, 110

*Xenesthis sp.*, 128

Yellow Flame Knee, 126
Yellow Legged Creeping Scorpion, 98
Yellow Toed Tarantula, 124
Yucatan Rust Rump, 122

Zebra Tarantula, 119-120